THE NAZI
CONNECTION

THE NAZI CONNECTION

Eugenics, American Racism, and German National Socialism

STEFAN KÜHL

New York   Oxford
OXFORD UNIVERSITY PRESS

Oxford University Press

Oxford New York
Athens Auckland Bangkok Bogotá Buenos Aires Cape Town
Chennai Dar es Salaam Delhi Florence Hong Kong Istanbul Karachi
Kolkata Kuala Lumpur Madrid Melbourne Mexico City Mumbai Nairobi
Paris São Paulo Shanghai Singapore Taipei Tokyo Toronto Warsaw

and associated companies in
Berlin Ibadan

Copyright © 1994 by Oxford University Press, Inc.

First published in 1994 by Oxford University Press, Inc.
198 Madison Avenue, New York, New York 10016

First issued as an Oxford University Press paperback, 2002

Oxford is a registered trademark of Oxford University Press

Library of Congress Cataloguing-in-Publication Data
Kühl, Stefan.
The Nazi connection: eugenics, American racism, and German national socialism / Stefan Kühl.
p. cm. Includes bibliographical references and index.
ISBN 978-0-19-514978-4
1. Eugenics—United States-History—20th century. 2. Eugenics—Government policy—Germany-
History-20th century. 3. Racism—Germany—History—20th century. 4. National socialism. I. Title
HQ755.5.U5K84 1994
363.9'2'09730904—dc20   93-17283

Printed in the United States of America
on acid-free paper

For Rebecca Jo

Preface

While researching the connection of German National Socialists to American eugenicists, I was able to view *Erbkrank* [Hereditary Defective], a Nazi race propaganda movie that was also used by the American eugenics movement for informing high school students about the need to sterilize mentally handicapped people. The movie showed mentally handicapped people living in a luxurious asylum near Berlin and contrasted their "atypicality" to the "saneness" of "hereditarily healthy" children who had to live in the slums of Germany's large cities. By stressing the "abnormality" of the handicapped people, this movie helped to pave the way for Nazi policies of mass sterilization and elimination of the mentally handicapped.

The years I spent working with such "atypical" and "abnormal" people at Protestant Youth in Munich was the impetus behind my decision to begin working on the history of mentally handicapped people under the Nazis. Without having met Karla Weber, Peter Schönauer, Elmar Wänke, Wolfgang Frisch, Franki Häusler, and many others, this book would not have been written. By revealing one of the darkest moments in the history of handicapped people, I hope to thank them for the many things they taught me.

Two of my history teachers deserve special acknowledgment. Horst Dieter Geetz at the Gymnasium of Quickborn introduced me to the multicausality of history, and helped me see the critical relevance of history to my work in the social sciences. My adviser at the University of Bielefeld, Gisela Bock, encouraged my interest in the development of scientific racism. Over the past four years she supported me in nearly every aspect of my work. She generously made time for lengthy discussions, shared many of her own sources, commented extensively on several of my papers, and helped me to gain financial assistance for the timely completion of this endeavor.

At an early stage of my work, Peter Weingart and Hans Walther Schmuhl from the University of Bielefeld helped me to clarify the outline of my project; they later commented on an early draft of this book. Michael Schwartz from the University of Münster was especially helpful in shaping my thinking about the concept of racism I used in this book. Paul Weindling from the University of Oxford and Peter Lindley at the University of Kent at Canterbury provided me with information and valuable comments. My research was made possible by two different organizations: The German Academic Exchange Service supported a one-year stay in the United States, and the Westfälisch-Lippische Universitätsgesellschaft agreed on short notice to provide the necessary resources for my archival work.

Many archivists and librarians in the United States, Germany, Great Britain, and France helped me uncover the explosive sources that show the extent and character of the relationship between Nazi and American scientists. I want to thank Martha Harrison of the American Philosophical Society in Philadelphia, Odessa Ofstad of the Pickler Memorial Library in Kirksville, Missouri, and Dr. Alan Burdock of the Milton S. Eisenhower Library at the Johns Hopkins University in Baltimore.

Several teachers and colleagues from my year at Johns Hopkins deserve my thanks. My adviser, Vernon Lidtke, patiently helped me organize my research and commented on several drafts of this book. Sharon Kingsland was exceptionally generous in sharing her impressive knowledge about eugenics and genetics in the United States; she encouraged me to rethink some aspects of my approach. I also want to thank Daniel Walkowitz and Leslie Reagan, both visiting professors at Johns Hopkins in 1991–1992, for commenting on a shorter version of this book.

Many American historians, notably Daniel Kevles, Garland Allen, Sheila Weiss, Robert Proctor, and Barry Mehler, introduced me to the latest research on American eugenics. Sheila Weiss and Robert Proctor commented on an early draft of this work. Barry Mehler provided insightful comments and spent several days discussing aspects of my research. He also generously shared many sources. Carl Degler and Robert Pois provided useful comments and convinced me to rewrite some details of an early draft.

Writing a study in a foreign language is always difficult, particularly under the burden of pressing deadlines and time constraints. At Oxford University Press, Nancy Lane, senior editor, and Edward Harcourt, editorial assistant, were enthusiastic about this project from the very beginning. Because they made themselves so available for me, this book was readied for the press in perhaps record time.

I also deeply appreciate the help of my friends and fellow graduate students in Baltimore. Colin Essamuah, Wolfgang Splitter, and Jürgen Wagner all gave editorial advice on various sections of an early draft. Tanya Kervokian looked over my German translations. Alisa Plant was always a great help in clarifying uncertainties about the use of language. During the final week of editing, Lynn Gorchov read and commented on the final version. My two dear roommates in Baltimore, Lori Bernstein and David Bernell, urged me to turn my research into a book and were unfailing sources of encouragement while I was writing the bulk of the manuscript.

More than anyone else, however, Rebecca Jo Plant participated in the genesis of this book. She painstakingly edited several drafts of the manuscript, helped me to clarify some of my ideas, and improved the style of the final draft. In the process, she convinced me that working on a fascinating subject can be, for a certain time, nearly the most important thing in life.

Paris   S. K.
May 1993

Contents

Introduction

My interest in the relations between German racial hygienists and American eugenicists emerged from my work in the archive of the largest Protestant institution for mentally handicapped and epileptic people in Germany, the von Bodelschwinghschen Anstalten in Bethel. My aim was to examine whether this famous German institution was the stronghold of resistance against the Nazi race program that it later publicly presented itself to have been. Fritz von Bodelschwingh, director during the entire period of Nazi rule, has become known as one of the main figures who resisted the extermination of mentally handicapped people during World War II. Sources located at Bethel and other institutions, however, led me to doubt the veracity of this interpretation.[1]

Like other leaders of the Protestant church in the late 1920s, Bodelschwingh was sympathetic to eugenic ideals, favoring, for example, sterilization of certain groups of the mentally handicapped. I was also impressed by the fact that, in discussions about eugenics at Bethel, the United States had played an important role as a model of a country in which eugenic sterilization and immigration legislation were at least to some degree successfully implemented. American supporters of Bethel mailed Bodelschwingh information about the progress of eugenics in the United States.

When I realized that I would not have access to the full range of Bethel's sources, I decided to turn my examination toward the role that the United States had played as a model for Germany, when under Hitler race improvement became a central component of German policies. However, as I surveyed German journals and newspapers from this time period, I was only secondarily impressed by references to the success of eugenics in the United States by Nazi race politicians. Indeed, I was more surprised by the broad coverage in Nazi propaganda of American scientists who expressed support for Germany's new policy of race improvement. When I turned to available secondary literature for more information about this group of American scientists, all active in the American eugenics movement, I was struck by the fact that their support for Nazi Germany had received little attention and tended to be obscured.

Historians writing in the 1960s and 1970s about the history of eugenics in the United States partially based their views on the self-portrayal of the American Eugenics Society.[2] After 1945 American eugenicists attempted to portray the relationship of American eugenics to Nazi Germany as distant and critical. The leadership of the American Eugenics Society after World War II either simply ignored their earlier relationship to Nazi Germany or falsely asserted that the Society had opposed Nazi race policies. They claimed that only an unimportant and marginal wing of the eugenics movement had reacted positively to mass sterilization, special support for "hereditarily valuable" couples, prohibition of miscegenation, and "euthanasia" in Nazi Germany.[3] They argued instead that in the 1930s eugenics in the United States became more scientifically oriented, while in Germany the Nazis "perverted" all science, and eugenics in particular, for the political purpose of improving the Nordic race.

In 1963, historian Mark Haller stated in the first monograph about eugenics in the United States that "between the mid-1920s and 1940 racism ceased to have scientific respectability, and as a result American eugenics and racism faced a parting of the ways." The idea of racial superiority survived only among "innumerable right-wing anti-semitic groups and among white supremacists" in the United States.[4]

Similarly, in the second important study of American eugenics, published in 1972, historian Kenneth M. Ludmerer distinguished between eugenicists critical of Nazi race policies and a small group of eugenicists who failed to see Nazi measures as a "perversion of the true eugenic ideal as seen by well-meaning men deeply concerned about mankind's genetic future."[5] This tendency to draw a sharp distinction between "true" eugenics and the perversion of eugenics by the Nazis continued to shape the historiography of eugenics throughout the 1970s. In a 1976 collection of essays about eugenics, the editor, Carl Bajema, strongly denied that American eugenics included "brutal racist evolutionary practices such as those of Nazi Germany."[6]

Subsequently, the approaches of Haller, Ludmerer, and Bajema were countered by attempts to show the involvement of American eugenicists in Nazi race policies. In 1977 historians Garland Allen and Barry Mehler revealed the connections of an especially prominent

American eugenicist, Harry H. Laughlin, to Nazi racial hygienists.[7] The same year, author Allan Chase published his comprehensive study, *The Legacy of Malthus: The Social Cost of Scientific Racism*, which reanalyzed the American eugenics movement's relationship to Nazi Germany. He claimed that it was the eugenics movement in the United States, and later in Nazi Germany, that "prompted state and national governments to make sterilization their weapon of choice against what the scientific racists called 'the menace of racial pollution.'"[8]

The 1980s witnessed new attempts to stress the similarities between the writings of American eugenicists and Nazi race policies. Scholars Thomas Shapiro and David Smith both dedicated short passages of their studies to discussing relations between German and American eugenicists after 1933.[9] Similarly, in a study concerning "psychiatric genocide" in Nazi Germany and the United States, Lanny Lapon, an activist in the anti-psychiatry movement, wrote about the similarity and continuity of racial ideology in both countries.[10]

"Mainstream" history, however, continued to underemphasize the Nazi connection of American eugenicists. In 1985 Daniel J. Kevles, historian at the California Institute of Technology in Pasadena, published *In the Name of Eugenics*, focusing on eugenics in the United States and Great Britain. Although Kevles claimed that his approach highlighted the influence of German racial hygiene on eugenics in both countries, he underestimated the importance of developments in Germany for the eugenics movement in the United States. In an otherwise excellent study, Kevles identified only two American eugenicists, Laughlin and Clarence G. Campbell, as supporters of Nazi Germany. In his view, by the mid–1930s "such racists constituted a rapidly diminishing minority, most of them isolated on the far political right."[11]

Historians only recently have begun to explore systematically the exact character of the relationship between American eugenicists and Nazi racial hygienists. In an in-depth study of the American Eugenics Society, Barry Mehler of Ferris State University in Big Rapids, Michigan, dedicated a chapter to comparing American and Nazi sterilization measures, drawing attention to their many similarities.[12]

Likewise, in a study about the history of coercive sterilization, scholar Stephen Trombley provided interesting new evidence in a chapter concerning Anglo–American cooperation with Nazi Germany.[13] In particular, he offered new insights into the role that California eugenicists played in supporting Nazi race policy.[14] Trombley, however, tended to view eugenicists without adequately distinguishing separate factions within the eugenics movement. The full range of responses to Nazi sterilization policy was therefore obscured.

The historiography of the American eugenics movement as a whole has suffered from a failure to use German sources, which provide a critical perspective on the interaction between German and American eugenicists. By drawing on such material, historian Robert Proctor succeeded in illustrating the significant influence on Nazi race policy of developments in the United States.[15] German historian Gisela Bock and scholars Peter Weingart, Jürgen Kroll, and Kurt Bayertz reached the same conclusion in their comprehensive studies about the German racial hygiene movement and the sterilization policy of Nazi Germany.[16]

Despite recent attempts to examine the support of Nazi race policy by non-German scientists and politicians, inquiries have remained restricted to exploring singular aspects of the Nazi connection to American eugenicists. Support for Nazi race policy often has been mentioned only in a very general sense, usually to illustrate the potential terror of eugenics. A more complete examination of the complex interaction between German and non-German eugenicists has been lacking. This lack of research concerning the collaboration between Nazi racial hygienists and their colleagues in other countries is surprising because the historiography of eugenics—in other countries as well as in Germany—has been strongly affected by the radicalization of eugenics in Germany after 1933. In other words, the Nazi uses of eugenics—including mass sterilization, the killing of handicapped persons, the murder of ethnic minorities, and the extermination of Jews—are always a silent presence in works about eugenics, even when not mentioned specifically. This influence can be detected by noting the manner in which historians have tended to construct their arguments. Historians have generally written about eugenics in two ways: Either they have emphasized similarities and continuities between eugenics and Nazi policies, or they have argued that certain aspects of eugenics should be distinguished from these policies.[17]

One reason why so little has been written about the interaction between Nazi racial hygienists and the eugenicists in other countries is the fact that the historiography has been limited by a national perspective. By focusing on eugenics as a national movement and a national science, historians have tended to overlook the issue of international collaboration. Although important recent studies acknowledge the international aspects of eugenics, transnational cooperation has not been adequately explored.[18]

My book seeks to correct this deficiency by providing such a perspective. I view my research as a contribution to the comparative study of eugenics, racial hygiene, and human genetics in different national contexts.[19] My focus on American eugenics and German racial hygiene opens up a new perspective for exploring the relationship of national eugenics movements to forms of state control, in particular the relationship of the German racial hygiene movement to the authoritarian Nazi system. As a scientifically and politically motivated attempt to improve the quality of humankind, eugenics existed under all types of governmental systems—democratic, fascist, and socialist—but the relationship between eugenics and state power clearly varied widely.[20] Therefore, an examination of the reaction of a eugenics movement in a Western democracy to the race policy of a totalitarian regime is further enhanced by an evaluation of the relationship of German racial hygienists to their own government. By analyzing how German racial hygienists were integrated into the Nazi government, we can gain insight into how social and political pressures shaped the behavior of a group of scientists. Adding American eugenicists, who were never under the authority of a totalitarian regime, into the analysis allows for an estimation as to what racial hygienists' and eugenicists' collaboration with the Nazis resulted from shared ideological principles. The thesis that the cooption of the racial hygiene movement in Germany was due to pressure imposed by the Nazis should be carefully scrutinized.[21]

The first two chapters deal with the period after 1945 and before 1933 in order to frame Nazi Germany within its historical context. In Chapter 1, I illustrate the present-day relevance of the historical relationship between American eugenicists and Nazi racial hygienists by exploring recent developments in scientific racism. Chapter 2 traces the development of the relationship of German and American eugenicists within the context of the international eugenics movement before 1933. In particular, I focus on how eugenic laws in the United States influenced discussions among German eugenicists in the Weimar Republic.

Chapter 3 explores how German racial hygienists and Nazi race politicians utilized the international eugenics movement for propaganda purposes after 1933. Chapter 4 deals with the shift from German racial hygienists viewing the United States as a role model to American eugenicists admiring Nazi race policies after the passage in 1933 of the

"Law on Preventing Hereditarily Ill Progeny." Chapter 5 focuses on trips that American eugenicists made to Nazi Germany in order to study the practical applications of Nazi race policies. In Chapter 6, I argue that even eugenicists who attempted to limit themselves to "purely" scientific contacts in Germany helped to stabilize the Nazi regime and that racism was the core of the ideology of both American and Nazi eugenicists.

Chapter 7 puts the development of the American eugenics movement into the context of the overall scientific community within the United States. Chapter 8 focuses on German racial hygienists and race politicians. I show how National Socialists used incentives to draw American eugenicists into supporting their propaganda strategy, and how acutely aware the Nazis were of the international reaction to their race policies. Chapter 9 analyzes the demise of relations between German racial hygienists and American eugenicists, beginning in the late 1930s and culminating with a complete break in 1941. However, I draw attention to the fact that, immediately after the war, German eugenicists asked scientists in the United States to support their reintegration into the international scientific community. The Conclusion summarizes continuity and discontinuity in the relationship between German and American eugenicists.

THE NAZI
CONNECTION

The "New" Scientific Racism

> Racism falsely claims that there is a scientific basis for arranging
> groups hierarchically in terms of psychological and cultural charac-
> teristics that are immutable and innate. In this way it seeks to make
> existing differences appear inviolable as a means of permanently
> maintaining current relations between groups.[1]
>
> UNESCO Statement on Race and Racial
> Prejudice, Paris, September 1967

The late 1980s witnessed a revival of public interest in scientific racism
on North American campuses. The media gave broad coverage to
research by scholars in the United States and Canada that attempted to
establish a scientific basis for classifying humans into "superior" and
"inferior" genetic groups.[2] For example, J. Philippe Rushton, professor at the University of Western Ontario, argued that whites and
Asians are generally more intelligent and family-oriented than were
blacks, while Johns Hopkins University sociology professor Robert
Gordon advanced the claim that the high crime rate among American
blacks correlated with their comparatively low intelligence level.

Roger Pearson's Justification

In 1991, anthropologist Roger Pearson jumped into the fray with what
was probably the most comprehensive defense of scientific racism in
the United States since 1945. In *Race, Intelligence and Bias in Academe,* Pearson denounced "the strong opposition by Marxists and
other Leftists" against research with racial implications.[3] He attacked
both academia and the media as bastions of politically motivated opposition to the pursuit of "objective" science.

Pearson's defense of research on questions of racial difference
stems from a long-term personal investment in such research. He has
been promoting the theory that the white race is endangered by inferior
genetic stock for more than thirty years. In the late 1950s, he helped to

found the Northern League and the journal *Northlander,* an initiative designed to foster the "interests and solidarity of all Teutonic Nations."[4] In 1978, he supported the World Anti-Communist League Meeting in Washington, D.C., which the *Washington Post* referred to as an assembly of the forces of "authoritarianism, neo-fascism, racial hierarchy, and anti-Semitism."[5]

Pearson has succeeded in combining such right-wing politics with a conventional academic career.[6] He served as director of the Council for Social and Economic Studies in Washington and today leads the Institute for the Study of Man in McLean, Virginia. In his fund-raising efforts for the Council for Social and Economic Studies, he proudly referred to a letter of support he received from President Ronald Reagan. On April 14, 1982, Reagan commended Pearson's "valuable service" and voiced appreciation for his "substantial contributions to promoting and upholding those ideals and principles that we value at home and abroad."[7]

The scientists whom Pearson views as threatened by "a powerful, politically motivated drive toward biological egalitarianism" include a group of American scientists who are conducting research that suggests that blacks, as a group, are genetically inferior to whites in intelligence.[8] Pearson treats the controversy between these scientists and their critics as part of a long history of "suppression of all realistic attitudes toward heredity and race" that followed the unfortunate demise of the eugenics movement.[9]

Eugenics, which Pearson defines in modern terms as "the practical application of genetic science toward the improvement of the genetic health of future generations," was a politically and scientifically influential movement in the first half of the twentieth century, particularly in Great Britain, the United States, and Germany. The word *eugenics* was originally coined by Francis Galton in 1883. He defined eugenics as the "science of improving the stock."[10] In his view, the eugenics movement should aim to give "the more suitable races or strains of blood a better chance of prevailing speedily over the less suitable."[11]

Pearson informs his readers that the original intention of eugenics was "clearly and unabashedly the goal of breeding a more gifted race."[12] According to Pearson, eugenicists believed that Europeans as well as other gifted races were already of distinguished genetic capability, but that "just as races differed genetically, so breeding groups of individuals within nations and regional populations might also differ genetically." Eugenicists concluded that "some individuals and breeding populations had genetically transmissible qualities, which were intellectually, physically, emotionally, and morally more desirable." Eugenicists employed two different approaches to improve the "national stock." "Negative eugenics," in Pearson's words, attempted "to free future generations from avoidable genetically transmitted handicaps." "Positive eugenics," on the other hand, sought to "raise the overall genetic quality of the nation by ensuring a superior birth rate among the genetically better-endowed."[13]

In *Race, Intelligence and Bias in Academe* Pearson attempts to disassociate eugenics from the shadow cast upon it by the extermination programs of Nazi Germany. After 1945, enthusiasm for research in eugenics and race questions had sharply declined as the full horror of Nazi uses of eugenics and race science became apparent. In the 1970s, when some academics again argued for the genetic inferiority of blacks and the preeminence of heredity over environmental influences, critics readily drew associations to Nazi ideology. Such scientists were accused of promoting fascist ideas.[14] Indeed, similarities between Nazi race ideology and racist research in the United States since 1945 have provided critics with a powerful means for attacking scientists with racist agendas. Pearson and his colleagues seem to recognize that it is mandatory to disassociate their research from association with Nazi Germany. In his introduction to Pearson's book, Hans J. Eysenck, a British psychologist known for his thesis that the white race is genetically more intelligent than the black race, attempts to turn the tables on his critics.[15] He claims that his attackers rely on force, not reason, and that the "the scattered troops of the 'New Left'" have adopted the "psychology of the fascists."[16]

Pioneer Fund's Financial Backing

Pearson's and Eysenck's outraged denials to accusations of Nazism, however, have to be considered in the light of the financial support behind Pearson's literary activities. Pearson's publications have been supported, in part, by the Pioneer Fund, a foundation whose early leadership had praised aspects of Nazi Germany's racial policies and which has, in more recent years, given financial support to controversial research into race and intelligence. Between January 1, 1986, and December 31, 1990, Pearson's Institute for the Study of Man received $214,000 from the Pioneer Fund, mostly for "literary activities."[17]

Harry H. Laughlin and Frederick Osborn, scientists who played a leading role in the American eugenics movement, and, as I will illus-

(AIDS) among blacks by pointing to their supposed reproduction strategies. Due to their lack of intelligence and social skills, Rushton and Bogaert argue, blacks can only compete with whites and Asians in the evolutionary process by maintaining a higher level of sexual activity. This could be proved, they asserted, by the fact that the penises and vaginas of blacks are larger on average, and that blacks have a higher premarital, marital, and extramarital intercourse frequency. The higher percentage of AIDS infections among blacks is therefore presented as the result of their genetically preeminent sexual behavior.[28] Rushton, who provided Pearson access to his personal files and published in Pearson's *The Mankind Quarterly,* has been heavily attacked in Canada and the United States. Pearson explains that the widespread protests against Rushton in Canada result from "the steady growth of immigrant power [in Canada] since the beginning of the present century."[29] Between 1986 and 1990, Rushton received more than $250,000 from the Pioneer Fund.

Robert Gordon is yet another protégé of the Pioneer Fund. He was not as creative as Rushton, but he was the author of a comprehensive collection of publications. Since the early 1970s, Gordon has promoted the notion that the differences in delinquency rates of blacks and whites are due to differences in their respective genetic constitutions.[30] Many of Gordon's academic publications repeat the thesis that a connection exists between race, inherited intelligence, and the tendency toward criminality.[31] In 1975 Gordon presented his thesis concerning the IQ-commensurability of racially specific delinquency rates.[32] In a paper presented to the annual meeting of the American Psychological Association in 1986, he repeated that intelligence is a more accurate determinant in accounting for the black–white differences in crime rates than is income, education, or occupation.[33]

Gordon, drawing on his status as a professor at Johns Hopkins University, defends colleagues who have been criticized for their research into race and intelligence. In *Race, Intelligence and Bias in Academe,* Pearson quotes Gordon's support for two such colleagues: Linda Gottfredson and Michael Levin. In 1990 he defended Gottfredson, a University of Delaware educational psychologist, against faculty members and students who protested her acceptance of Pioneer Fund money. Gordon called the Fund one of "the last sources of private support that courageously operates at all in this intellectually taboo arena."[34] In a letter defending Michael Levin, of City College of the City University of New York, he wrote:

If our nation is to deal rationally with the awkward but extremely conse-
quential fact of group differences in various mental abilities, which are
the rule rather than the exception, and not tear itself apart instead in an
ideological frenzy, future leaders of all races are going to have to learn
about those differences and how to ponder their implications in a civil
and mutually respectful manner.[35]

Gordon received $124,000 from the Pioneer Fund between 1986 and
1990.

In the 1980s, the largest share of Pioneer Fund money went to
support controversial "twin studies" at the University of Minnesota;
over $500,000 was awarded between 1986 and 1990 alone.[36] At the
Minnesota Center for Twin and Adoption Research, psychologists
study twins who were raised apart to determine how much of behavior
is grounded in heredity. Psychologist Thomas J. Bouchard and his
colleagues follow Jensen, Rushton, and Gordon only in that they argue
for the predominance of inherited over environmental influences. Statements about differences between races are not an aspect of the Minnesota project. Their goal is to prove that tendencies toward religiosity,
political radicalism, or tolerance toward sexual minorities are to a large
extent inherited, as are preferences and capacities for certain profes-
sions. Bouchard and his colleagues conclude, based on their findings,
that the possibility of influencing intelligence and learning abilities is
slim.[37]

While not racist in itself, this thesis has been adopted by Pearson
and his colleagues as important proof that genetic factors set the potential limits of human behavior, while the influence of environmental
circumstances is determined by heredity. Based on the research at the
University of Minnesota, which he praises "as one of the great successes of modern American science," Pearson draws conclusions
about differences among races. For example, from the result that a
"conservative, an authoritarian, or a liberal nature, as well as rebel-
liousness, and aggressiveness, even political preferences" have heritable biological roots, Pearson hopes to extrapolate conclusions about
racial differences in personality as well as IQ.[38]

The support of the Pioneer Fund is not limited to Jensen, Shockley, Pearson, Rushton, Gordon, and the Minnesota Project. The list of
other recipients of Pioneer Fund grants reads partly like a "Who's
Who" of scientific and political racism in the United States, Canada,
Great Britain, and Ireland. Recipients include the American Immigra-

tion Control Federation, the Foundation of Human Understanding, Richard Lynn, professor of psychology at the University of Ulster, Eysenck's Institute of Psychiatry at the University of London, and Seymour Itzkoff of Smith College.[39]

"Nazi Methods" or "Nazi Ideology"

In the conflict between those who receive Pioneer Fund money and the opponents of the Fund, both sides have accused each other of using "Nazi methods" or espousing "Nazi ideology." For example, Gordon accused Mehler of acting like the former Nazi minister of propaganda, Joseph Goebbels, in his criticism of Linda Gottfredson:

> Goebbels would admire Mehler's technique of first inflaming emotions by calculated references to Hitler and the Klan and then promptly channeling those emotions against academics doing research that he opposes, but which he cannot refute through normal scholarship.[40]

In the same article, Gordon implied that those who criticized Jensen, Gottfredson, and himself would bring fascism to America, only under another name.[41]

Similarly, Eysenck has compared the behavior of many of his colleagues to that of Germans under the Nazi government. Although recognizing the correctness of Eysenck's and Jensen's theses, they were confronted by "hostile students" and therefore refused to extend support outside of private conversations. Eysenck concluded that it was in just such a manner that many Germans become anti-Semites "under duress."[42] Eysenck's biographer, H. B. Gibson, commented that the Nazis had been defeated in war, but "anyone with Eysenck's intelligence and grasp of reality knew that the execution of a few psychopaths" solved little in historical terms. In Eysenck's eyes, according to Gibson, "the most powerful modern heirs of the Nazis were the various extreme political groups who often identified themselves as 'communists' or 'Marxists.'"[43]

Some academics have charged that researchers studying purported racial differences in intelligence are promoting the same ideology that dominated Nazi Germany. Mehler has argued that the Pioneer Fund, in addition to providing financial assistance to research that stands in the tradition of Nazi race ideology, was actually created by men who supported Hitler's racial ideology. Confronted with this charge, and aware of the stakes involved, the president of the Pioneer Fund, attorney Harry Weyher, denied all connections between the founding fathers of his institution and the leaders of Nazi Germany. In a letter to the *American Jewish World,* Weyher asserted that "it is highly unlikely that two such prominent men" as Laughlin and Osborn could have supported Hitler without public knowledge.[44]

In the conflict about scientific racism, the word *Nazi* has degenerated into a term to be used in any situation to discredit the opponent. By providing detailed evidence about the relationship between American eugenicists and Nazi Germany, I hope to ground references to Nazi Germany in the recent controversies about scientific racism on a historically secure basis. The evidence that I present about the history of the Pioneer Fund between 1937 and 1945 and the enthusiasm of its founders for Nazi Germany is not intended to be the only argument against scientific racism. In disputes with scientists active in race research, it is clearly not enough to cry "Nazi." The development, however, of science in general and scientific racism in particular needs to be seen within its proper historical context. The Nazi connection with American scientists and its continuity as manifested in the Pioneer Fund can help us understand Nazi race ideology and the results and implications of present-day race research as part of a shared history of scientific racism.

German–American Relations within the International Eugenics Movement before 1933

The forceful and decisive North American does not consider the traditional moral code and does not consider the individual in order to implement what he thinks is right. After he recognizes the importance of heredity in determining mental and physical traits for the entire population, he does not hesitate to proceed from theoretical reflection to energetic practical action and to enact legislation which will lead to ennoblement of the race.[1]

German eugenicist Feilchenfeld in 1913

In an interview for the *Berliner Tageblatt*, Alfred Ploetz, the German founder of the science of racial hygiene, discussed his experience at the first International Congress for Eugenics held in London in 1912. Ploetz, who served as president of the International Society for Racial Hygiene, described the United States as a bold leader in the realm of eugenics.[2] His comments foreshadowed the development of a relationship between German and American eugenicists that was grounded in an emerging international community of scientists dedicated to the goal of race improvement.

The Early International Connection

The groundwork for the first major international meeting of eugenicists was laid at a meeting of racial hygienists during the 1911 International Hygiene Exhibition in Dresden. Organized by the International Society for Racial Hygiene, a group founded in 1907 and dominated almost exclusively by German racial hygienists, this meeting brought together eugenicists from Germany, the Netherlands, Czechoslovakia, Great Britain, Austria, Sweden, Denmark, and the United States. The purpose of the meeting was to foster international ties and to publicly present the results of the rising new science.[3]

The International Congress of 1912 was longer and more comprehensive than the Dresden meeting, drawing over 300 participants from Europe and the United States. Leonard Darwin, son of the famous evolution theorist Charles Darwin and head of the British Eugenics Education Society, the official sponsor of the Congress, presided. Many famous scientists and other prominent individuals served as vice-presidents, including the American inventor, Alexander Graham Bell; Charles B. Davenport, director of the Eugenics Record Office in Cold Spring Harbor, located in Long Island, New York; Charles W. Eliot, president of Harvard University; and David Starr Jordan, president of Stanford University. Ploetz and Max von Gruber, professor of hygiene in Munich, served as German vice-presidents. Great Britain was represented by Winston Churchill, then secretary of state for Home Affairs, and William Collins, vice-chancellor of the University of London. Lucien March, director of the Institute for Statistics in Paris, and Edmond Perrier, director of the Museum for Natural History in Paris, acted as vice-presidents from France, while August Forel, a famous psychiatrist from Zurich, represented Switzerland.

The Congress was separated into four sections. The first section dealt with the question of heredity, primarily the physical aspects of heredity and the issue of miscegenation. The second section concentrated on the influence of eugenics on sociological and historical research. The third section treated the impact of eugenics on legislation and social practices. The last section considered the practical applications of eugenic principles. In the final section, participants discussed how to prevent procreation of the "unfit" through segregation and sterilization, and how to encourage procreation of the "fit" by promoting eugenic ideals.

The invitation to the Congress declared its purpose as:

> [T]o make more widely known the results of the investigations of those factors which are making for racial improvement or decay; to discuss to what extent existing knowledge warrants legislative action; and to organize the cooperation of existing societies and workers by the formation of an International Committee or otherwise.[4]

The Congress succeeded in fulfilling its stated goals, particularly regarding the mission of international organization. The London Congress strengthened existing informal contacts between eugenicists of different countries and led to the creation of the Permanent International Commission of Eugenics.

Despite the fact that the International Commission promised to provide German racial hygienists with important contacts to eugenicists in Great Britain and the United States, its founding represented a defeat for Ploetz and his colleagues. Ploetz had hoped to strengthen and extend the influence of the International Society for Racial Hygiene by integrating more non-German eugenicists into his organization. However, only the Scandinavian eugenicists supported a merger with the International Society for Racial Hygiene, which would have endorsed German leadership. Ploetz was forced to accept British domination of the emerging international organization for eugenics. Although international meetings of eugenicists ceased during World War I, the foundation for transnational cooperation had been laid.

American eugenicists enjoyed a strong standing among their foreign colleagues. European eugenicists admired the success of their American counterparts in influencing eugenics legislation and gaining extensive financial support for the American eugenics movement. The German racial hygiene movement followed the development of the American eugenics movement closely. During World War I, the Society for Racial Hygiene in Berlin distributed a public flyer extolling "the dedication with which Americans sponsor research in the field of racial hygiene and with which they translate theoretical knowledge into practice." The flyer referred to a donation of several million dollars by a widow of a railway magnate in support of the eugenics laboratory in Cold Spring Harbor. Also mentioned was a foundation established in 1915 following a eugenics conference in Battle Creek, Michigan, which provided $300,000 for conferences and exhibitions in the field of eugenics. According to the flyer, this financial support facilitated intensive research in the field of heredity, including Alexander Graham Bell's extensive studies on longevity.

The flyer also claimed that American farmers believed that racial hygiene was the most important question of the century, and praised the funding of state commissions that attempted to awaken the eugenic consciousness of the nation. It applauded the "fantastic" control of immigration through restrictive legislation, as well as laws in twelve states designed to prevent procreation of "inferior families." The Society for Racial Hygiene concluded that Americans recognized the critical importance of race improvement and were eager to adopt measures to further this goal. The flyer ended with the rhetorical question: "Can we have any doubts that the Americans will reach their aim—the stabilization and improvement of the strength of the people?"[5]

The reason German racial hygienists in general and Berlin racial hygienists in particular were so well informed about the situation of eugenics in North America was due in part to one of the most active members of the Berlin society. Géza von Hoffmann, who spent several years as the Austrian vice-consulate in California, regularly informed his German colleagues and the German public about eugenic developments in the United States.[6] In 1913, he published a book, *Die Rassenhygiene in den Vereinigten Staaten von Nordamerika* [*Racial Hygiene in the United States of North America*], which later became one of the standard works of the early eugenics movement. After an introduction that sketched the scientific basis of eugenics, he reported on the widespread acceptance of eugenic ideals in the United States. He claimed that Galton's hope that eugenics would become "the religion of the future" was being realized in the United States.[7]

The theories of evolution and decay [Entartung], the importance of heredity, and the possibility of race improvement—in short, the ideas of Darwin, Mendel, and Galton—were penetrating American scientific thought and social life. As evidence, Hoffmann quoted Woodrow Wilson's presidential address in which he claimed "that the whole nation has awakened to and recognizes the extraordinary importance of the science of human heredity, as well as its application to the ennoblement of the human family."[8] The United States, Hoffmann argued, recognized that limited reproduction of "blue-blooded" Yankees would lead to "race suicide." This phrase, coined by sociologist Edward A. Ross in 1901 and later adopted by Theodore Roosevelt, expressed the fear that "inferior" segments of the population were gaining power.[9] Hoffmann pointed out that federal and state agencies had established commissions to examine how eugenics could be used in state policy and had provided eugenic research with financial support. He dedicated an entire chapter to describing marriage restrictions applied to "unfit" and "unsocial" elements of American society. He reported that marriages of "feebleminded" persons were restricted in the majority of states, but complained that the measures were not implemented as rigorously as the laws in thirty-two states that prohibited marriage and sexual intercourse between blacks and whites.[10]

Hoffmann dedicated the largest section of his book to sterilization legislation, which, in his opinion, represented the "easiest measure to prevent the reproduction of inferior people."[11] He informed his reader that the first eugenic sterilization performed in the United States occurred in Indiana in 1899, without a legal basis.[12] In 1907, the doctor who performed the procedure convinced Indiana legislators to enact a law allowing for sterilization of the mentally handicapped. In 1909, California and Connecticut enacted similar measures, followed in 1911 by Nevada, Iowa, and New Jersey, and, in 1912, New York. In 1913, Kansas, Michigan, North Dakota, and Oregon also passed sterilization laws.

Hoffmann's final chapter addressed the eugenic orientation of American immigration restrictions. He explained that American eugenicists demanded that selection be both individually and racially based. The *"Homo Europaeus,* the Germanic and Nordic" type, served as the model of racial superiority. Hoffmann quoted American eugenicist Charles Woodruff as stating, "It is clear that the types of human beings from northwest Europe are our best citizens and have, therefore, to be conserved."[13]

German and English eugenicists praised the importance of Hoffmann's information about eugenics in the United States, since it allowed European eugenicists insight into events on the other side of the Atlantic. The only criticism came from Fritz Lenz, coeditor of the major German journal for racial hygiene, who argued that Hoffmann's account seemed to exaggerate the success of eugenics in the United States.[14] Lenz reproached Hoffmann for overestimating the effectiveness of sterilization laws and marriage restrictions, which had only limited influence as long as the most "capable" segments of the American population continued to practice birth control. Lenz claimed that the "extreme dominance of the ladies" accounted for the low birth rates among Anglo-Americans.[15] He argued that it was much more critical to support the procreation of "hereditarily worthy" people than it was to concentrate on hindering the reproduction of "inferiors." He admitted that the negative eugenic measures in the United States were more advanced than they were in Germany, but pointed out discrepancies between the laws and actual practice. He argued that this lack of enforcement was not surprising in a nation governed by an "extremely democratic administration," in which even administrators were elected by the masses.[16]

The different positions voiced by Hoffmann and Lenz reveal the conflicting perceptions of American eugenic measures held by German eugenicists prior to World War I. German eugenicists normally acknowledged the leading role of the United States in implementing eugenic legislation, but they criticized American policies as haphazard and poorly enforced. Until the late 1910s, Géza von Hoffmann remained the primary link between German and American eugenicists, although contact became increasingly difficult with the outbreak of World War I. Hoffmann gradually grew less optimistic regarding the future of eugenics in the United States. He claimed that rash actions, the lack of a powerful bureaucratic system, and the peculiarity of the American Constitution were partially responsible for getting sterilization laws passed, but also contributed to their poor enforcement.[17] In 1914, he reported in the journal of the International Society for Racial Hygiene, the *Archiv für Rassen- und Gesellschaftsbiologie* (*ARGB*), on a proposal of the American Genetic Association, which was so "unbelievably radical" that he was unsure as to whether or not to take it seriously. The Commission of the American Genetic Association, headed by Harry H. Laughlin, proposed that the lowest 10 percent of the American population be sterilized. This extreme measure, never seriously considered for state legislation, was intended to "eradicate" the "inferior" members of the society over a time period spanning two generations.[18] Despite his doubts concerning the feasibility of such measures, Hoffmann praised the proposal for accurately illustrating the extent to which sterilization needed to be implemented.

After World War I: Reintegrating German Racial Hygienists

World War I strained international relations among eugenicists. The Second International Congress of Eugenics was postponed, and the Permanent Committee ceased meeting until October 1919. During this meeting, which was held in London, eugenicists from Belgium, Great Britain, Australia, Denmark, France, Italy, Norway, and the United States agreed to hold a Second International Congress for Eugenics in 1920 or 1921.[19] The Congress took place in New York City in September 1921, without German participation. In the aftermath of war, such formal cooperation was out of the question, but the international eugenics community had nonetheless already started to reintegrate individual German racial hygienists into their ranks. Charles B. Davenport, the main organizer of the Congress, expressed his regrets to Agnes Bluhm, one of the early German racial hygienists, and apologetically explained to Alfred Ploetz that "international complications have prevented formal invitations to the International Eugenics Congress in New York City."[20] He expressed his hope that such complications would be resolved before the next conference.[21]

Indeed, Davenport played the central role in reintegrating German racial hygienists into the international eugenics movement. Acting on the initiative of two famous Scandinavian eugenicists, Davenport, as the newly elected president of the Permanent International Committee on Eugenics, used his influence to grant German racial hygienists a stronger position within the movement. His gesture, however, was rebuked by his German colleagues. For example, in 1923 Erwin Baur, a famous German geneticist, and eugenicist Fritz Lenz turned down an invitation to join the meeting of the international organization. Baur, who with Lenz and Eugen Fischer authored the main eugenics textbook in Germany, thanked Davenport for his overture, but declared that German racial hygienists would not sit on a committee with French and Belgian eugenicists as long as French and Belgian troops occupied the Ruhr.[22] In the same vein, Lenz argued that as long as parts of Germany were occupied by foreign troops, "there is no time for international congresses."[23] Instead, he proposed strengthening bilateral exchanges between German and American eugenicists.[24] The following year, however, German racial hygienists agreed to send delegates to a meeting of the international organization. In October 1924, the annual general meeting of the German Society for Racial Hygiene (the new name for the prior International Society for Racial Hygiene) agreed to send Alfred Ploetz and its president, Otto Krohne, as representatives to the next international eugenics meeting, but demanded that German be accepted as a conference language, and that neither Brussels nor Paris be chosen as the conference site.[25]

By the time Germany began to rejoin the international movement in 1925, relations between German and American eugenicists were already restored. Fritz Lenz assumed Géza von Hoffmann's role as the main link between the movements. He established positive relations with Laughlin and Davenport at the Eugenics Record Office in Cold Spring Harbor and cooperated closely with Paul Popenoe, an important eugenic figure on the West Coast. In a 1924 article about the German racial hygiene movement translated by Popenoe, Lenz stated that there were virtually no differences between the position of eugenicists in the United States and Germany. He confessed that Germany lagged behind in terms of legislation, which he explained by stating that "the Germans are more disposed toward scientific investigation than toward practical statesmanship." Nevertheless, he was confident that if eugenic education proceeded, legislation would naturally follow.[26] Looking back in 1934, historian Reinhold Müller concurred with Lenz's view. In an article for *Ziel and Weg,* Müller wrote:

Racial hygiene in Germany remained until 1926 a purely academic and scientific movement. It was the Americans who busied themselves earnestly about the subject. Through massive investigations, they proved (with impeccable precision) Galton's thesis that qualities of the mind are as heritable as qualities of the body; they also showed that these mental qualities are inherited according to the very same laws as those of the body.[27]

Popenoe, Lenz's colleague on the other side of the Atlantic, regularly reported on American developments in the journal of the German racial hygiene movement.[28] Roswell H. Johnson, coauthor of *Applied Eugenics* with Popenoe, explained that the advancement of eugenics was nowhere greater than it was in the United States and in Germany. This, Johnson argued, was the result of a wider base of interest that, in turn, was the result of more general higher education in both countries than anywhere else.[29]

Underlying the close working relationship between America and Germany was the extensive financial support of American foundations for the establishment of eugenic research in Germany. The main supporter was the Rockefeller Foundation in New York. It financed the research of German racial hygienist Agnes Bluhm on heredity and alcoholism in early 1920. Following a European tour by a Rockefeller official in December 1926, the Foundation began supporting other German eugenicists, including Hermann Poll, Alfred Grotjahn, and Hans Nachtsheim. The Rockefeller Foundation played the central role in establishing and sponsoring major eugenic institutes in Germany, including the Kaiser Wilhelm Institute for Psychiatry and the Kaiser Wilhelm Institute for Anthropology, Eugenics, and Human Heredity.[30]

In 1918, German psychiatrist Emil Kraepelin founded the Institute for Psychiatry in Munich, which was taken over by the Kaiser Wilhelm Society in 1924. The Department for Genealogy and Demography was headed by Ernst Rüdin, later director of the Institute for Psychiatry. This department—the core of the Institute—concentrated on locating the genetic and neurological basis of traits such as criminal propensity and mental disease. In 1928, the Rockefeller Foundation donated $325,000 for the construction of a new building. The funding of the Institute in Munich was a model that other American sponsors followed. Ironically, the Institute continued to be supported by the money of the Jewish philanthropist James Loeb until 1940.

The actual building of the Kaiser Wilhelm Institute for Anthropology, Eugenics, and Human Heredity in Berlin was also partially funded by money from the Rockefeller Foundation. At the opening celebration in 1927, Davenport, still president of the International Federation of Eugenic Organizations (IFEO), delivered a speech in the name of the international eugenics movement. The Institute concentrated on a comprehensive project on racial variation as indicated by blood groups, and on twin studies, coordinated by Otmar Freiherr von Verschuer.[31] When severe financial problems threatened to close the Institute during the early years of the Depression, the Rockefeller Foundation kept it afloat. At several points, the Institute director, Eugen Fischer, met with representatives of the Foundation. In March 1932, he wrote to the European bureau of the Foundation in Paris, requesting support for six additional research projects.[32] Two months later, the Rockefeller Foundation answered affirmatively. The Foundation continued to support German eugenicists even after the National Socialists had gained control over German science.

By 1930 the United States and Germany had surpassed Great Britain as the leading forces of the international eugenics movement. At this time, Davenport was succeeded as president of the IFEO by Ernst Rüdin, director of the Kaiser Wilhelm Institute for Psychiatry in Munich. The IFEO committees, established in the late 1920s, were influenced mainly by scientists from Germany and the United States. The Committee on Human Heredity based its work on the studies of Rüdin and his colleague at the Kaiser Wilhelm Institute, as well as on Hans Luxemburger's study of the heredity of psychopathology, schizophrenia, and manic depressive insanity, Eugen Fischer's work on the genetics of tubercular diathesis, and von Verschuer's studies of identical twins.[33] The Committee on Race Crossing was jointly led by Davenport and Fischer. Both agreed that "the contrast between the slight scientific activity in the field of hybrid investigation and the vast extent of race crossing in almost all parts of the earth" was unfortunate.[34]

The Eugenics Record Office and the Station for Experimental Evolution in Cold Spring Harbor and the Kaiser Wilhelm Institute in Berlin together prepared a questionnaire in English, German, French, Spanish, and Dutch that was distributed to 1,000 physicians, missionaries, and consulates to collect information about miscegenation in different areas of the world. Fritz Lenz, chairman of the Committee for Race Crossing, insisted at the conference of the IFEO on September 27, 1928, that the Federation as a whole should be more engaged in supporting their work on race mixing. Fischer suggested that "Jew—

Gentile crosses," available in most European countries, could be excellent subject material, while "bastard twins" also promised to provide a wealth of data.[35] The Committee on Race Psychiatry, chaired by Rüdin, attempted to examine the relationship between race and insanity. Members of the committee suspected that "inferior" races were more likely to show a higher rate of mental retardation, schizophrenia, and manic depression than the white race.[36]

The Third International Congress of Eugenics, held in 1932 in New York, was a setback for Germany in terms of extending German influence within the international eugenics movement. Important racial hygienists like Fischer and Ploetz were unable to attend the Congress because of economic difficulties brought on by the Depression. The theme of the Congress was "A Decade of Progress in Eugenics."[37] Indeed, the 1920s were an era of progress for eugenicists in many respects, despite the fact that some former supporters of the eugenics movement, such as Raymond Pearl, Herbert S. Jennings, and Hermann Muller, had grown critical of key figures in the eugenics movement and had ceased participating in eugenics organizations. In a note to American newspapers, the organization committee of the Congress claimed that "to a greater extent than ever before the evolution of the lower organisms is under our control."[38] Future possibilities in the field of heredity, claimed the American *Eugenic News,* would call for eugenicists to collaborate with investigators in the fields of history, anthropology, physiology, psychology, medicine, statistics, plant and animal genetics, and other closely related sciences.[39]

The "decade of progress" referred both to advances within so-called pure eugenics as well as "applied eugenics," which meant increased educational and legislative activity. A growing number of college professors in the fields of psychology, biology, and sociology were offering courses in eugenics. The sterilization movement in different nations of the world had also advanced. In 1928, the Swiss Canton Vaud passed a law allowing for sterilization of mentally handicapped persons if health administrators foresaw the danger that the individuals in question would produce degenerate offspring. Denmark followed one year later with a similar law. By 1928, the Eugenics Society in Great Britain had initiated a comprehensive campaign for voluntary sterilization, which led to discussions in the British Parliament, though legislation was never actually enacted. *Eugenic News* concluded that "eugenics as a 'long time investment in family-stocks' is making substantial headway."[40]

American Influence on Germany before 1933

Germany, too, witnessed the rise of a strong campaign for sterilization. In 1932, a committee of the German Medical Association and the Prussian Health Council [Landesgesundheitsrat] proposed to limit medical care for handicapped people and to implement legislation that would allow for voluntary sterilization. In discussions of the Prussian Health Council, Benno Chajes, urologist and socialist member of the Prussian Parliament, drew on existing sterilization laws in twenty-four states of the United States, as well as a Swiss law, in order to illustrate the benefits of sterilization legislation.[41] Indeed, the entire German sterilization discussion prior to the implementation of the Law on Preventing Hereditarily Ill Progeny, passed on July 14, 1933, was strongly influenced by American models.

The first attempts to implement a sterilization law in Germany stemmed from the one-man initiatives of Gerhard Boeters, a district physician in Zwickau, Saxony. In May 1923, Boeters sent a report to the government of Saxony in which he demanded compulsory sterilization for the hereditarily blind and deaf, the mentally handicapped, the mentally ill, sexual "perverts," and fathers with two or more illegitimate children. He published a model law, the so-called Lex Zwickau, in several regional newspapers and in the medical press.[42] Boeters referred directly to the experience in the United States, stating,

> In a cultured nation of the first order—the United States of America, that which we strive toward [sterilization legislation] was introduced and tested long ago. It is all so clear and simple.[43]

Americans of German origin, he believed, would be especially interested in his plan. When writing to the State Department, he asked the government to support the distribution of his model law to fifty German newspapers in America.[44]

In 1923, the Reich Health Office, directed by Franz Bumm, faced legal, religious, scientific, and political barriers to enacting a sterilization law. Opponents claimed that racial hygiene had not provided conclusive proof that sterilization could effectively reduce the number of mental and physical "inferiors." Furthermore, the turbulent political atmosphere in Germany in 1923 did not provide a favorable setting for a legislative act that would have led to serious disagreement in scientific, political, and economic circles. Nevertheless, the Reich Health Office decided to initiate an inquiry in the United States, based on Géza von

Hoffmann's book, regarding the legal and scientific basis of steriliza-tion.[45] In the fall of 1923, the German embassy and consulates in the United States began an extensive examination, which revealed that the implementation of sterilization laws in several states had ceased, and that "sterilization in the United States compared to the first decade of the century does not play such an important role."[46]

Although Boeters was initially isolated regarding the sterilization issue, his initiative garnered interest among hygienists, psychiatrists, and lawyers.[47] In 1927, three and a half years after the State Depart-ment survey, the Social Democratic faction in the Prussian Parliament unsuccessfully filed a petition urging the government to again collect material about the eugenic results of sterilizations in North America.[48] The Social Democrats' initiative signaled the importance of the United States as a role model for Germany, while also indicating that interest in such legislation extended to the left of the political spectrum.[49]

After 1925, scientific and medical literature about sterilization regularly referred to the United States. Robert Gaupp, professor at the University of Tübingen, reported in an influential pamphlet concerning the sterilization of "mentally and morally ill and inferiors" that, con-trary to the position of the Reich Health Office, sterilizations in the United States "were increasing quickly." Actual figures support this observation. In the thirteen years from 1907 to the beginning of 1920, 3,233 persons were sterilized, while in the four years from 1921 to 1924, 2,689 persons were sterilized—a much higher annual rate than in the 1910s. The average rate of 200–600 sterilizations per year before 1930 shot up in the 1930s to 2,000–4,000 sterilizations per year.[50] Although in favor of eugenics, Gaupp was cautious in promoting com-pulsory sterilization. He claimed it was ironic that, in contrast to the United States—"the country of freedom"—the "right of self-determination" in Germany was too strong to allow for the adoption of eugenic principles.[51]

The late 1920s witnessed a rapid increase of interest in sterilization questions and consequently in the experiences of the United States. In 1929, Harry H. Laughlin, the assistant director of the Cold Spring Harbor Laboratories, published an article about legislative develop-ments in the United States in the influential *ARGB*. The article was based on a talk he had presented at the meeting of the IFEO in Munich one year before. Laughlin provided German readers with detailed in-formation about the status of sterilization laws in twenty-three states. He claimed that eugenic sterilization was no longer considered a radical method in the United States: "It has been proven that sterilization is necessary to the well being of the state.'' However, he also stressed that laws alone were insufficient and needed to be enacted in conjunction with eugenic education, marriage restrictions, and other measures. Most importantly, the ''prohibition of procreation for certain members of degenerate tribes'' needed to be accompanied by special support for marriages deemed hereditarily valuable. He closed:

> The racial hygienist as a biologist regards the development of eugenic sterilization as the effort of the state ''organism'' to get rid of the burden of its degenerate members.[52]

Two books from 1929 provided German racial hygienists with extensive material about the situation in the United States. A study on sterilization in California by eugenicists Eugene S. Gosney and Paul Popenoe was translated into German only one year after it appeared in the United States.[53] Felix Tietze from the Austrian League for Regeneration and Heredity claimed that ''nobody who is working on the question of eugenic sterilization could neglect this study.''[54]

German sterilization expert Otto Kankeleit also published a book based on experiences in the United States. He referred to the 1927 Supreme Court decision that ruled in favor of the constitutionality of compulsory sterilization. Kankeleit referred to Laughlin's studies when he demanded that sterilization of ''inferior'' women should have priority. In the opinion of both eugenicists, the number of ''degenerate'' individuals depended mainly on the number of ''degenerate'' women: ''Therefore the sterilization of the degenerate woman is more important than that of the man.''[55]

The importance of the United States for German eugenicists was revealed by the allusions in nearly every German medical dissertation about sterilization in the United States as the first country to enforce comprehensive eugenics legislation.[56] These dissertations often referred to literature by Géza von Hoffmann, Hans W. Maier, and Laughlin.[57] One explanation given for the United States' leading role in eugenics was that racial conflicts in the United States had forced the white population early on to employ a systematic program of race improvement.[58] The dissertations normally supported the compulsory character of American sterilization, but were critical concerning the lack of enforcement.[59]

Such admiration—limited only by doubts about some aspects of the sterilization laws—also extended beyond sterilization laws and marriage restrictions. In particular, the American Immigration Restriction Act of 1924 was applauded by German racial hygienists. Hans F. K. Günther, a famous German race anthropologist, praised the measure for its joint approach of prohibiting both degenerate individuals and entire ethnic groups from entering the United States. In an article entitled "The Nordic Ideal" Bavarian Health Inspector Walter Schultz wrote that German racial hygienists should learn from the United States how to restrict the influx of Jews and eastern and southern Europeans. He took the fact that the immigration law had drastically reduced annual immigration as evidence that "racial policy and thinking has become much more popular than in other countries."[60] One other important German figure, in a famous book from 1924, was full of praise for the fact that the Immigration Restriction Act excluded "undesirables" on the basis of hereditary illness and race. His name was Adolf Hitler; the book was *Mein Kampf*.[61]

3

The International Context: The Support of Nazi Race Policy through the International Eugenics Movement

To that great leader, Adolf Hitler![1]

American eugenicist Clarence G. Campbell
at a reception during the 1935 International
Population Congress in Berlin

The International Federation of Eugenic Organizations

In the summer of 1934, one and a half years after the Nazis came to power in Germany, the International Federation of Eugenic Organizations (IFEO), meeting in Zurich, passed a resolution to which Nazi propaganda frequently referred in order to illustrate the international acceptance of their race policies. In this unanimously passed resolution, sent to the prime ministers of all the major Western powers, the IFEO stated that, despite all differences in political and social outlooks, the organization was "united by the deep conviction that eugenic research and practice is of the highest and most urgent importance for the existence of all civilized countries." It recommended that all governments "make themselves acquainted with the problems of heredity, population studies, and eugenics." It stated that eugenic principles should be adopted as state policies "for the good of their nations . . . with suitable regional modifications."[2]

German racial hygienists and Nazi race politicians viewed this resolution as confirmation of German and American dominance in the eugenics movement and as international approval of the 1933 German sterilization law. Although the resolution did not refer directly to Germany, its adoption was seen as an achievement for National Socialists in gaining international acceptance of their policies.[3] Nazi racial hygienist Heinz Kürten, who led a Committee for the Implementation of the National Revolution with the goal of forcing Jews out of medical positions in Germany, explained that the conference had shown eugenicists from all over the world that the implementation of comprehensive eugenics measures in Nazi Germany represented an important step in global eugenics.[4] Likewise, at a reception for foreign diplomats and the international press on March 21, 1935, Walter Gross, director of the National Socialist Party's Office for Education on Population Policy and Racial Welfare [Aufklärungsamt für Bevölkerungspolitik und Rassenpflege], which was soon renamed the Racial Policy Office [Rassenpolitische Amt der NSDAP], referred to this resolution as of central importance to an assessment of Nazi race policies.[5]

Prior to the conference, the Nazis were aware of how important the event could be in gaining scientific and political recognition of their race policies and countering the generally negative responses in foreign newspapers toward their new sterilization law. Leading figures of national eugenics movements attended, such as Jon Alfred Mjöen of Norway, Morris Steggerda of the Eugenics Research Association of the United States, George Schreiber of France, and Hans W. Maier, director of the Psychiatric Clinic in Zurich. Eugenicists from Great Britain, East India, Denmark, Poland, British Borneo, and Austria also participated.

The German delegation was the largest at the conference, and the leader of the conference was Munich racial hygienist Ernst Rüdin.[6] As chairman of the IFEO Committee on Race Psychiatry, Rüdin spoke about the relationship between mental retardation and race. Lothar Loeffler, an influential figure regarding sterilization administration, urged eugenicists in his presentation not to hesitate to draw political conclusions from their scientific research. Freiherr von Verschuer, the leading researcher of twins, presented a talk about the use of such studies in research on mental retardation. He was accompanied by colleagues Ernst Rodenwaldt from Frankfurt and Lothar Tirala from Munich. In addition, leading figures of the Nazi administration also participated in the conference. One such figure was race politician Karl Astel, who reported on the practical adoption of eugenics in the German state of Thuringia.

Of most interest to those reporting at the conference, however, was a talk by Falk Ruttke, a lawyer and member of both the S.S. (Hitler's elite guard) and the Committee for Population and Race Policies in the Reich Ministry of the Interior. Ruttke was one of the primary people involved in the construction of Nazi race ideology. At the conference, he reported how Germany's "unfavorable, not to say disastrous" population situation had improved since Hitler had come to power. Before 1933, according to Ruttke, Germany's declining birth rate "left only the dependent part of the community rising in numbers." Since then, he claimed, knowledge of genetic laws had been invoked to create a "healthy race."

Ruttke proceeded to outline all the steps the Nazis had taken, beginning with a measure designed to combat unemployment, which he viewed as leading to family breakdown. The Law to Reduce Unemployment, enacted July 1, 1933, attempted to replace women workers with men through the implementation of state-funded work and through occupational training for the unemployed. The next step was to foster procreation through marriage subsidies to young persons of "good stock." The Decree for the Granting of Marriage Loans, passed July 1, 1933, allowed funding to non-Jewish couples free of mental or physical illness.[7] This measure to support "valuable" couples was accompanied by attempts to eliminate "inferior" members of the society. The Law on Preventing Hereditarily Ill Progeny, passed July 14, 1933, allowed for the sterilization of persons with different mental and physical afflictions. The Law against Dangerous Habitual Criminals, enacted on November 24, 1933, allowed for the sterilization and castration of criminals.

Another important step taken by the Nazis to improve the quantity and quality of the German people was to provide special support to rural settlements. The Hereditary Homestead Law, passed September 29, 1933, and the Law for the New Formation of the German Farmerstock, passed July 14, 1933, provided more than 100,000 new homesteads for families of "good stock" and subsidized "hereditarily valuable" farmers. Ruttke quoted Reich Minister for Agriculture and Reich Leader for Farmers Richard Walther Darré, who claimed that farmers were "the most valuable blood source" of the German people. Implementation of these various eugenic measures was guaranteed through the centralization of the public health administration, following the passage of the Law for the Unification of Health Administration on July 3, 1934. In addition to overseeing the coordination of public health measures, the purpose of this administration was mainly to provide support for "hereditary and racial care."[8]

The German race policies in general and Ruttke's speech in particular played an important role in determining how the Conference as a whole was evaluated by eugenicists of different countries. The American *Journal of Heredity* reported on Ruttke's speech as illustrating the eugenics foundation of the new Nazi state.[9]

At the 1936 IFEO conference in Scheveningen, the Netherlands,

German racial hygienists again constituted the largest group, and Nazi race policies again dominated the part of the conference that dealt with "applied eugenics." Fifteen delegates from Germany attended, as compared with five from the Netherlands and three each from the United States and England. Austria, Denmark, and France all sent two delegates, while Sweden, Norway, Estonia, and Latvia were represented by one delegate. Reports were presented concerning new research on the inheritance of mental disorders, methods for research in the psychology of inheritance, the mutation rate in plants, animals, and humans, and statistics of selection. Reports about eugenic policies in different nations also played an important role. Charles M. Goethe, president of the Eugenics Research Association, explained to the European eugenicists that because of the "low qualitative composition" of certain strains in the American population, the United States had taken strong measures to prevent the further admission of undesirable immigrants and to purge the existing population. Caroline H. Robinson, a member of the board of directors of the Eugenics Research Association, informed the delegates that approximately two-thirds of America's female college graduates did not marry. Ernst Rüdin, Falk Ruttke, and Karl Astel discussed the Sterilization Law in Germany. Only one participant, Dutch geneticist G. P. Frets, criticized the compulsory character of the German law.[10]

The minutes of the meeting indicate that the participants appreciated the information presented by German racial hygienists. The fact that Ernst Rüdin, the past IFEO president and director of the Kaiser Wilhelm Institute for Psychiatry, was serving as a chief adviser to the Nazi government was viewed as a "great opportunity." The participants paid special attention to Ruttke's and Astel's talks.[11] Ruttke reported on the "progress" that Nazi race policies had made since the previous conference two years earlier. Since 1934, he reported, tax laws in Germany were based on racial ideology. The sterilization law, the most sensational of the Nazi measures for race improvement, had been further extended. Amendment laws passed on June 26, 1935, and February 4, 1936, legalized abortion if the pregnant woman had already been singled out for sterilization.[12] Furthermore, a decree dated February 26, 1936, allowed for the sterilization of women by radiation.

To illustrate his account, Ruttke distributed brochures that included the texts of the Law on Preventing Hereditarily Ill Progeny, the two amendment laws, and five decrees, translated into English, French, Spanish, and Italian. Ruttke also devoted substantial time to a discussion of marriage restrictions. The Law for the Protection of Heredity, passed on October 18, 1935, prohibited marriage between "healthy" and mentally retarded persons. Ruttke stressed the important role of physicians in the improvement of the German race. The Reich Decree for the Medical Profession, passed on December 13, 1935, declared it the duty of the German medical profession to protect the health of both individuals and the German people. As a whole, doctors were deemed responsible for the "stabilization and improvement of health, hereditary value, and the race of the German people." The race legislation, Ruttke explained, was accompanied by comprehensive race propaganda and measures to improve environmental factors. He concluded by stressing that the legislation would be unsuccessful if the National Socialists failed to convince the population of the need to protect its hereditary value:

> Hereditary traits are not only given to us, but carry a moral obligation to pursue the highest biological development possible. This not only calls for work on behalf of the volk, into which the individual is born and with which he is connected through blood ties, but also on behalf of all humankind. This is thus extremely important work toward the maintenance of peace.[13]

Germany's scientific press and the Nazi mass media reported extensively on the 1936 Conference.[14] The *Völkischer Beobachter*, mouthpiece of the Nazi government, stated that, despite different world views, the conference accepted the "absolutely leading position of Germany in genetic research and in practical measures in the area of racial welfare." The *Völkischer Beobachter* concluded that "leading racial hygienists of nearly all civilized nations have agreed with the German position and accepted the correctness of the measures implemented in Germany."[15]

The 1934 and 1936 conferences reveal the domination of the IFEO by German racial hygienists and their foreign supporters in other countries. The Nazi bureaucracy and German racial hygienists agreed that winning approval of the IFEO was crucial for gaining international acceptance of their race policy. Their strategy was to dominate the international conferences, to support only eugenicists friendly to National Socialism as leaders of the IFEO, and to provide as many German organizations as possible with access to the Federation. In 1935 and 1936, three new German institutes dedicated to the study of racial hygiene and human genetics joined the organization.

Ernst Rodenwaldt, representative of the Institute for Hygiene of the University of Heidelberg, author of a study on race crossing, and coeditor of the central German scientific journal for racial hygiene, wrote to the German minister for education that "it is self-evident that the participation of German racial hygienists in this early international organization is a necessity."[16] The Nazi administration concurred. In a letter to the Reich Ministry for Propaganda, an administrator of the Reich Ministry of the Interior stated that his department favored a "quantitatively and qualitatively excellent representation of German scholars" at the meetings of the International Federation.[17] In the late 1930s, the Federation virtually depended on Nazi Germany; the extent of German influence can be seen by the fact that in 1939 the Federation accepted German-incorporated Vienna as the site of the Fourth International Congress for Racial Hygiene and Eugenics.[18]

The 1935 International Congress for Population Science in Berlin

The 1935 International Congress for Population Science in Berlin marked the apex of international support of Nazi race policies and represented a great success for the Nazi race propaganda machine. This Congress assembled prominent eugenicists, anthropologists, population scientists, and geneticists from all over the world. German racial hygienists constituted the largest group of participants, delivering 59 of the 126 presentations.

The speeches presented at the Congress represented the entire spectrum of eugenics and population science. Charles Close, the British president of the International Union for the Scientific Investigation of Population, spoke about population trends in Great Britain. He called for a reduction in the rate of population growth and for an improvement in the overall quality of the society. Two other people from Great Britain voiced support for this view. Marie C. Stopes, president of the Society for Birth Control and Progress of Race, stressed the qualitative results of births as central for the survival of human beings. C.B.S. Hodson, secretary of the IFEO, spoke of "the biological worth of the stable family." Scientists from Switzerland, Austria, Latvia, the United States, Germany, France, Hungary, Spain, Italy, and India reported about population movements in their respective countries. A large section was dedicated to "racial hygiene as common task for civilized nations." B. Sekla, from the Czechoslovakian Eugenic Society, claimed that in his nation "inferior" members were reproducing themselves much faster than their "superior" counterparts.

One of the few individuals to raise any criticism at the Congress was a French participant who attacked the scientific basis of sterilization laws. He was angrily refuted by German and non-German eugenicists like Mjöen, Hodson, and Rüdin. Questions about race typologies were discussed by numerous delegates: French race anthropologist Réné Martial; leader of the eugenics movement in Italy, Corrado Gini; Swiss eugenicist Otto Schlaginhaufen; and two German race anthropologists, Egon Freiherr von Eickstedt and Albert Harrasser. An important section was dedicated to practical racial hygiene. Arthur Gütt, ministerial director in the health department in the Reich Ministry of the Interior, director of the Reich Committee for Public Health Service [Reichsausschusses für Volksgesundheitsdienst], and chief author of the sterilization law, introduced the section by using a stock phrase of the Nazi government: "General need goes before individual need" [Gemeinnutz geht vor Eigennutz]. He was followed by Rüdin, Verschuer, Astel, Ruttke, and Ploetz.

The 1935 Congress had been initiated in 1931 by the International Union for the Scientific Investigation of Population Problems (IUSIPP), an international scientific organization in the field of population sciences closely connected to the IFEO, its eugenic counterpart. IUSIPP did not alter plans to hold the Congress in Berlin, although leading figures of IUSIPP recognized that the Nazi government would use the Congress for its own purposes.

The president of IUSIPP, Raymond Pearl, remained committed to Berlin as the conference site, although he feared that population science would become politicized.[19] Pearl, professor at Johns Hopkins University, was initially a eugenics enthusiast who became critical of the movement in the 1920s and turned to population science. Pearl honored the president of the Population Conference, Eugen Fischer, as a distinguished and broad-minded scientist.[20] He accepted Fischer's invitation to serve as vice-president as "a great honor."

As it turned out, Pearl was unable to attend the conference.[21] Instead, two other Americans served as vice presidents at the Berlin conference in 1935: Harry H. Laughlin and Clarence G. Campbell. Laughlin could not go to Berlin, but nevertheless accepted the honorary position of vice-president, contributed a paper entitled "Further Studies on the History and Legal Development of Eugenic Sterilization in the United States," and sent an exhibit, consisting of twelve illustrative charts and publications.[22] Included were testimonies given before the United States Congress in 1924 that illustrated how the United States had founded its immigration policy on biological principles.[23]

Clarence G. Campbell, who served as the senior representative of the American eugenics movement in Berlin, delivered a lecture on "Biological Postulates of Population Study." After praising a number of non-German eugenicists, he proceeded:

> It is from a synthesis of the work of all such men that the leader of the German nation, Adolf Hitler, ably supported by the Minister of Interior, Dr. Frick, and guided by the nation's anthropologists, its eugenicists, and its social philosophers, has been able to construct a comprehensive race policy of population development and improvement that promises to be epochal in racial history. It sets the pattern which other nations and other racial groups must follow, if they do not wish to fall behind in their racial quality, in their racial accomplishment, and in their prospect of survival.[24]

German race hygienists were conscious of the important role Campbell played in rallying support for Nazi race policy. In his closing speech about "race and achievement [Rasse und Leistung]," the leitmotif of the conference, Fischer made special reference to the "excellent remarks of Mr. Campbell."[25]

An "impressive result of the Congress," according to German propaganda in scientific journals and the mass media, was "the fact that well-known researchers from all over the world welcomed as effective and promising the line taken by the National Socialist government . . . and that they thought it is necessary that all countries' governments should follow the Germans on this path."[26]

The Nazi press showed a special interest in Campbell. The *Völkischer Beobachter* reported him as stating:

> The Third Reich under the guidance of racially conscious men has established a comprehensive race policy of population development and race improvement [Volksaufartung] based on the knowledge of eugenic science. This fact will secure Germany a place in the history of races.[27]

In an interview with the *Berliner Börsenzeitung*, Campbell praised the Berlin Congress as an extraordinarily important meeting that had succeeded in convincing representatives from all over the world that race biology should be at the center of every population policy.[28]

After his return to the United States, Campbell attempted to garner support among his American colleagues for the race policies in Nazi Germany. "Anti-Nazi propaganda with which all countries have been flooded," he lamented, "has gone far to obscure the correct understanding and the great importance of the German race policy."[29] In an article for *Eugenic News,* the official organ of the Eugenic Research Association, the Galton Society, and the American Eugenics Society, Campbell claimed that Nazi race policies had gained "the enthusiastic support and cooperation of practically the entire German nation." He argued that evidence of public support of "racially valuable families" could already be seen in Germany's increasing birth rate. "Where American families desire another motor-car, when they can afford it, German families desire another child."[30]

Der Erbarzt, supplement to the *Deutsche Ärzteblatt,* edited by the *Deutsche Ärztevereinsbund* and the *Verband der Ärzte Deutschlands (Hartmannbund),* reprinted Campbell's article in a translated version as a sign of international support for Nazi race policies.[31] *Der Angriff,* a widely read Nazi newspaper, quoted Campbell's article in *Eugenic News* as evidence of support by a well-known scientist publishing in an internationally accepted journal. The Nazi journal *NSK* noted Campbell's article as proof of acceptance of race-oriented measures on the other side of the Atlantic.[32] Under the headline "Amerikanische Forscher fordern Anwendung des Sterilisationsgesetzes in der ganzen Welt" [American Researchers Demand Application of the Sterilization Law Throughout the World], another German newspaper reported on a speech by Campbell before the Canadian Club in Toronto. On that occasion, Campbell claimed that it would be necessary to sterilize 10 percent of the population, lest the world suffer from racial degeneration.[33]

Campbell's statements on German race policy were exceptional in their enthusiasm; he virtually collaborated with the Nazi propaganda machine. Shortly after his comments appeared in the American press, German papers reprinted them to illustrate international scientific support for German race policies. Indeed, Campbell was the most frequently cited non-German scientist in the Nazi press. Campbell's statements, however, represent only the most extreme example of collaboration. Nazi race policies were widely accepted in some international scientific circles and by American eugenicists in particular. The reason eugenicists supported a policy that was criticized by many highly esteemed scientists requires further explanation.[34]

The Consensus between Nazi Race Politicians and Eugenicists in Other Countries

Nazi eugenics measures—including sterilization, marriage restrictions for unwanted members of society, and their exclusion from government subsidies, which were reserved for people defined as "valuable"—corresponded with the goals of eugenicists all over the world. Eugenicists understood Nazi policies as the direct realization of their scientific goals and political demands. In 1934, Leon F. Whitney, secretary of the American Eugenics Society, expressed his admiration for the German sterilization law. "Many far-sighted men and women in both England and America," he stated, "have long been working earnestly toward something very like what Hitler has now made compulsory."[35]

Eugenicists recognized that Hitler's steps toward improving the "German race" represented both the implementation of their practical proposals and, more importantly, the adoption of their basic ideology. Regardless of nationality or affiliation within the eugenics movement, all eugenicists urged governments to be "eugenically minded" in matters of political programs and social organization.[36] The world, they argued, should operate according to biological principles.[37] Nazism implemented this kind of thinking on an unprecedented scale. At the 1936 IFEO meeting, Falk Ruttke explained how the German government had designed all measures of racial welfare according to the scientific results of eugenics. To him, this represented the consistent "adaptation of biological knowledge to statesmanship."[38] Rudolf Hess, the deputy leader of the Nazi Party, expressed the same thought by employing a popular Nazi expression first coined by Fritz Lenz in 1931 (in even simpler terms): "National Socialism is nothing but applied biology."[39]

The appeal of National Socialism for eugenicists was strong. For the first time, their ideas became the basis for the organization of a whole state. *Eugenic News* announced that "nowhere else than in Germany are the findings of genetics rigorously applied to the improvement of the race."[40] In the other important eugenics journal in the United States, the *Journal of Heredity*, Paul Popenoe, a California member of the board of directors of the American Eugenics Society, praised Hitler for basing "his hopes of biological regeneration solidly on the application of biological principles of human society."[41]

From Disciple to Model: Sterilization in Germany and the United States

> The Germans are beating us at our own game.[1]
>
> Joseph S. DeJarnette, member of the
> Virginia sterilization movement

Eugenicists in the United States were the strongest foreign supporters of Nazi race policies. Other national eugenics movements, such as the one in Great Britain, were relatively critical of Nazi Germany. The *Rassenpolitische Auslandskorrespondenz,* the main German observer of foreign positions toward Nazi race policies, published eleven reports on eugenic activities within the United States. Four of these articles dealt with the support of the American eugenics movement for Nazi policies. No other country played such a prominent role in Nazi propaganda.[2]

The Influence of American Eugenics on Nazi Race Policy

The Nazi administration referred to the "model U.S." as playing an important role in shaping its own race policy. Otto Wagener, head of the Nazi Party's Economic Policy Office from 1931 to 1933, wrote about Hitler's personal interest regarding eugenic developments in the United States. He claimed that Hitler said:

> Now that we know the laws of heredity, it is possible to a large extent to prevent unhealthy and severely handicapped beings from coming into the world. I have studied with great interest the laws of several American states concerning prevention of reproduction by people whose progeny would, in all probability, be of no value or be injurious to the racial stock. I'm sure that occasionally mistakes occur as a result. But the possibility of excess and error is still no proof of the incorrectness of these laws.[3]

37

In 1935, the *Rassenpolitische Auslandskorrespondenz* stated that, in terms of eugenics, Germany was a "good disciple of other civilized societies."[4] In 1939, the *Archive für Rassen- und Gesellschaftsbiologie* claimed that the United States had "achieved something great" since the passage of its first sterilization measures.[5] Likewise, *Volk und Rasse* referred favorably to United States Supreme Court decisions that legitimized compulsory sterilization in 1916 and again in 1927. In order to prevent "being swamped with incompetents," the court argued:

> It is better for all the world if instead of waiting to execute degenerate offspring for crime, or to let them starve for their imbecility, society can prevent those who are manifestly unfit from continuing their kind. The principle that sustained compulsory vaccination is broad enough to cover the cutting of the Fallopian tubes.[6]

Although sterilization in the United States was more limited than it was in Germany, German racial hygienists emphasized that sterilization practices in some parts of the United States were more radical than were those in Nazi Germany. German economist Paul Heinz Besselmann explained the early acceptance of "such drastic measures" in the United States by pointing to the willingness of American politicians to implement "radical" laws.[7] However, German racial hygienists opposed the policy adopted by some American states of using sterilization as punishment and criticized the arbitrary way in which state governments enforced sterilization measures. They proudly referred to their own elaborate decision-making process, implemented by special "courts for hereditary health" in Nazi Germany.

American immigration laws designed to keep out people with hereditary diseases and citizens from non-Nordic countries won special approval in Germany. German economist H. H. von Schneidewind claimed that the aim of such policies was the preservation of Nordic blood. He was impressed by the influential role that the eugenic studies of Lothrop Stoddard and Madison Grant had played in shaping the thinking and policies of the Harding administration.[8]

In February 1934, Hans F. K. Günther, race anthropologist and a special protégé of the Nazis, explained to his audience in a crowded hall at the University of Munich that it was remarkable that "American immigration laws were accepted by the overwhelming majority, although the United States appeared the most liberal country of the world." He referred to Grant and Stoddard as the "spiritual fathers" of immigration legislation and proposed that the laws serve as a model for

Germany.[9] Nazi racial hygienists were especially impressed by the way in which American immigration policy combined eugenic and ethnic selection.[10]

American eugenicists were conscious and proud of their impact on legislation in Nazi Germany. They recognized that the German Law on Preventing Hereditarily Ill Progeny was influenced by the California sterilization law and designed after the Model Eugenic Sterilization Law, developed by Harry Laughlin in 1922. The German law followed Laughlin's proposal in terms of basic guidelines, but it was slightly more moderate.[11] Laughlin called for the sterilization of the mentally retarded, insane, criminal, epileptic, inebriate, diseased, blind, deaf, deformed, and economically dependent. The German law demanded sterilization in cases of mental retardation, schizophrenia, manic-depressive insanity, inherited epilepsy, Huntington's chorea, hereditary blindness, deafness, and malformation. It also allowed for the sterilization of alcoholics under a separate category. Both laws delegated the power of decision making to a special court. *Eugenic News* commented that "to one versed in the history of eugenic sterilization in America, the text of the German statute reads almost like the American model sterilization law."[12]

Access to information regarding legal and medical aspects of sterilization in the United States was one reason why the Nazis were able to pass the sterilization law in Germany within a mere six months after its takeover. In a letter to the Reich Ministry of the Interior in Berlin, the prime minister of Thuringen, Fritz Sauckel, explained that German legislators had to rely on reports from foreign countries due to a lack of experience in their own country.[13] Likewise, Ruttke claimed that before the German sterilization law was passed, German experts had carefully examined the experiences of foreign countries regarding sterilization.[14] The German sterilization law was the first such legislation to be based on a systematic analysis of experiences and discussions abroad.[15]

The Nazi Reception of American Degeneration Studies

In addition to learning from the practical and legal experiences of sterilization in the United States, the Nazis also drew upon research conducted in the United States after 1870. The first family eugenics study, based on an examination in thirteen jails in Ulster County, New York, was carried out in the mid-1870s by William L. Dugdale, a New York merchant and prison reformer. Dugdale examined four families with blood ties in order to prove that pauperism was a hereditary trait. According to his study, a frontiersman named Max Juke married a degenerate wife and produced an astonishingly large line of "white trash." Of the 709 descendants Dugdale claimed to have located, he identified 181 prostitutes, 106 illegitimate births, and 142 beggars. Furthermore, he stated that 64 descendents were housed at public expense and 70 had been convicted of crimes, 7 for murder. Dugdale estimated that this family alone cost the taxpayers of New York over $1.3 million between 1730 and 1874.

Dugdale's study inspired a wave of research about "degenerate" families. For example, Reverend Oscar McCulloch conducted a study of the "social degradation of the Indian tribe of Ishmael," and the dean of the graduate school at the University of Kansas, Frank W. Blackmar, focused on the "Smoky Pilgrims" in 1897. After 1904, Cold Spring Harbor Laboratories coordinated several family studies. In 1907 Charles Davenport's wife, Gertrude C. Davenport, published a report about the "Zero Family," based on records of a Swiss insane asylum. The first such study based on research of the Eugenics Record Office was Charles Davenport's report on "hill folks."

Without doubt, the most famous of the family studies was that of the Kallikaks from 1912, conducted by Henry Herbert Goddard and his field worker, Elizabeth Kite. Goddard was director of research at a school for mentally retarded children in Vineland, New Jersey, and the translator and publisher of Alfred Binet's intelligence test in the United States. *Kallikaks*—from the Greek words *kallos* [beauty] and *kakos* [bad]—was a pseudonym for the descendents of a soldier who served in the Revolutionary Army. In 1776, Martin Kallikak had sex with a nameless "feebleminded tavern wench," who bore an illegitimate boy. The descendents of the bastard numbered 480, including 143 mentally retarded, 36 illegimate births, 33 sexual deviants, 24 alcoholics, 3 epileptics, 82 infant deaths, and 3 criminals. Martin Kallikak reformed after leaving the army and married a "respectable girl of good family." Through that union another line of descendents arose of a radically different character. Among the 496 descendents, only three were found to be "somewhat degenerate." All of his legitimate children married into good families, including descendents of colonial governors, signers of the Declaration of Independence, and the founders of Princeton University. Goddard wrote that in the *kallos* family and its collateral branches, "we find nothing but good representative citizenship." In his opinion, the Kallikak study proved that "feeblemindness is hereditary and transmitted as surely as any other characteristic."[16]

Eugenicists used the scientific evidence of these family studies to support their calls for sterilization as a means of stopping the exponential reproduction of "degenerates." For example, during a 1927 Supreme Court session, eugenicists referred to the Kallikak study in their testimonies as proof of the hereditary character of mental retardation and the need for sterilization.[17] Likewise, in 1932, at the Third International Congress of Eugenics, Theodore Russell, from the Essex County Mental Hygiene Clinic in New Jersey, referred to the family studies conducted in the United States in a speech entitled "Selective Sterilization for Race Culture." Russell assumed that there were few in his audience

> [W]ho have not read the descriptions of the trail of crime, murder, pauperism, prostitution, illegitimacy, and incest which is found in the history of the famous Jukes and Kallikak families. It was demonstrated that the main factor in these ignoble family histories was mental deficiency. It would have cost but $150 to have sterilized the original couples, to cut off the seemingly endless social sores resulting wherever members of these families have settled. Yet the actual cost of relief alone of only one of these families was estimated at over $2,000,000 in 1916, as there were at that time 2,000 members of that socially unworthy clan.[18]

National Socialists enthusiastically adopted the stories of the Jukes and Kallikaks in order to legitimize their own sterilization program. The first German edition of Goddard's book about the Kallikaks was published in 1914. The second edition appeared in November 1933 in a special issue of *Friedrich Mann's Pädagogisches Magazin*. In the introduction, translator Karl Wilker clarified the impact of the Kallikak study:

> Questions which were only cautiously touched upon by Henry Herbert Goddard at that time . . . have resulted in the law for the prevention of sick or ill offspring dated the 14th of July, 1933. These questions have become generally interesting and significant. Just how significant the problem of genetic inheritance is, perhaps no example shows so clearly as the Kallikak family.[19]

The *Neues Volk* commented that the new edition showed that "feeble-mindedness . . . is the best fecund soil for every form of crime."[20] The Kallikak study also played an important role at the German Exhibition for Hereditary Care in 1935. The *Wochenblatt Sachsen Anhalt* reported on a controversy between a doubtful visitor and a Nazi doctor. After several questions about the character of the Kallikak study, the visitor asked:

— And who has carried out all this research?
— This examination was initiated and directed by the American Professor Goddard. There is even a book about it.
— Yes, Doctor, one last word. All the cripples and idiots here—all are due to the same cause?
— Yes. There is only one answer: heredity.[21]

The doctor, however, did not mention that Goddard had distanced himself from his study and its political implications as early as 1928.[22]

In addition to the Kallikak study, a prominent family study in Nazi race propaganda was Dugdale's earlier examination of the Jukes. The *Zeitschrift für Rassenkunde* praised Dugdale's study as the first to prove the hereditary character of "inferiority." Referring to a follow-up study by Arthur Estabrook from the Eugenics Record Office, it claimed that it had subsequently been possible to locate up to 2,820 of these "American criminals." The social cost of the Juke clan had risen to over $1.3 million. The *Zeitschrift für Rassenkunde* concluded that the high fertility rate and the hereditary nature of criminality had been proven in the United States, and that the American government would surely start to think about the millions of dollars being wasted upon these families of criminals.[23]

The California Sterilization Experience

Especially important for the German law were the detailed analyses of sterilization measures adopted in California. Popenoe and his colleagues in the California sterilization movement regularly informed German racial hygienists before and after 1933 about new developments in California, the state where nearly half of all sterilizations in the United States were performed. In 1935, a representative of the American Committee on Maternal Health visiting Nazi Germany detected the influence of the California experience on the German Law on Preventing Hereditarily Ill Progeny. After discussions with members of the Nazi administration and with judges of the Hereditary Health Courts, she concluded that:

> The leaders in the German sterilization movement state repeatedly that their legislation was formulated only after careful study of the California

experiment as reported by Mr. Gosney and Dr. Popenoe. It would have been impossible, they say, to undertake such a venture involving some 1 million people without drawing heavily upon previous experience elsewhere.[24]

An essential basis for the development of the German sterilization law was a study by Popenoe and Eugene S. Gosney, president of the primary eugenics organization in California, the Human Betterment Foundation. Originally published in 1929, *Sterilization for Human Betterment* appeared in a German edition the following year.[25] The authors, after examining 6,000 sterilized persons, reported that the operations had led to a decline in sex crimes. This was intended to counter one of the main arguments of social reformers who opposed sterilization in part because they believed sterilized women were more likely to become prostitutes. However, Popenoe and Gosney could positively ascertain such a result in only one case. Referring to this study, the *Völkischer Beobachter* claimed that the example of California illustrated the "beneficial effects" of sterilization laws: Nearly one-half of the sterilized "feebleminded" women were married or had at one point been married.[26]

In the second half of 1933 and the beginning of 1934, the Human Betterment Foundation mailed an influential pamphlet detailing California's experiences to German racial hygienists and Nazi administrators responsible for the enforcement of the German law. The brochure claimed that sterilization served to protect the sterilized person, his or her family, and society at large. It closed by asserting that people were becoming increasingly convinced that a nation that asked its able citizens to risk their lives in times of war was entitled to demand a much smaller sacrifice from its incapable citizens in times of peace.[27] In a cover letter to an administrator of the Innere Mission, a social welfare organization of the German Protestant Church that was particularly active on eugenic issues, Gosney applauded the fact that "with the adoption of a eugenic law by Germany, more than 150 million civilized people are now living under such a law."

Hans Harmsen, chief propagandist of the Protestant Church for eugenic measures, justified the Nazi sterilization law by pointing to the brochure and California's prior experience.[28] Two of the main Nazi politicians promoting sterilization, Arthur Gütt and Herbert Linden, also used the pamphlet to further the cause of the German sterilization law. Linden, an influential member of the Health Department at the Reich Ministry of the Interior and later a chief organizer of the Nazi extermination of more than 100,000 mentally handicapped people, referred to the report of the Human Betterment Foundation in his speech before the Committee for Population and Racial Policies. Citing the experiences of the United States, he claimed that opposition to the Law on Preventing Hereditarily Ill Progeny was due to anticipation that the law would be enforced with unprecedented thoroughness, and not due to the content of the law.[29]

During the 1930s, the Human Betterment Foundation and the California Branch of the American Eugenics Society remained important sources of information for Nazi Germany. Popenoe, the leading figure of both societies, received special attention. In an article about him and the California eugenics movement, *Der Erbarzt* portrayed Popenoe as a eugenicist of international stature and argued that the journal of the Human Betterment Foundation enjoyed influence throughout the American continent.[30]

Concerned about the public acceptance of their own sterilization law, German propaganda reported in 1936 that, according to a survey of the Human Betterment Foundation, the overwhelming majority of Californians supported sterilization laws. According to the survey, more than 90 percent of people with some knowledge of sterilization approved such measures; the only critics were those who were ignorant of the issues. The Nazi journal *NSK* viewed this survey as evidence that the more information people had about sterilization, the more likely it was that they were to support it. The newspaper concluded:

> [E]ducating the people on the character of sterilization cannot be insistently and comprehensively enough. This holds not only for California, where sterilization by the state has taken place for a whole generation, but for Germany as well, which, with its exemplary "Law on Preventing Hereditarily Ill Progeny," ranks above all other nations.[31]

Up to the late 1930s, German scientific journals and Nazi propaganda reported new publications, developments, and demands from the California eugenics movement.[32]

American Support for the German Sterilization Law

In view of such recognition, it is not surprising that Popenoe, Gosney, and the California eugenics movement as a whole strongly supported the Nazi sterilization law. In 1934, the California eugenics movement organized the presentation of an exhibition of the Reich's eugenics program. The exhibition was shown during the annual meeting of the

American Public Health Association and was opened to the public at the Los Angeles County Museum. The newsletter of the Southern California Branch of the American Eugenics Society promoted the exhibition:

> It portrays the general eugenics program of the Nazi government, giving special attention to the need for sterilization. Those who have seen this exhibit say it is the finest thing of the kind that has ever been produced. Take the opportunity to see this while it is in Los Angeles. Tell your friends about it.[33]

Popenoe viewed the German sterilization law as the fulfillment of principles developed by the California movement. He remarked that "since the Nazis came into full power, changes have been so frequent that it has been difficult to keep track of them."[34] He announced that the German sterilization law, which became effective on January 1, 1934, encompassed "the largest number of persons who had ever been included in the scope of such legislation at any one time."[35] He called the German law well conceived and argued that it could be considered superior to the sterilization laws of most American states. In a letter to L. C. Dunn, a critic of Hitler's race policy, Popenoe defended the German sterilization law:

> The law that has been adopted is not a half-baked and hasty improvisation of the Hitler regime, but is the product of many years of consideration by the best specialists in Germany . . . I must say that my impression is, from a close following of the situation in the German scientific press, rather favorable.[36]

In a scientific evaluation of sterilization laws in different countries, Popenoe identified favorable trends during the first three years following the application of the German sterilization law.[37]

The example of California reveals the critical role that the transfer of knowledge about medical, scientific, and political aspects of sterilization played in the formulation of Nazi sterilization legislation. However, support of the Nazi sterilization law was not limited to California eugenicists. Other state organizations for eugenics and sterilization were also enthusiastic supporters. In a letter to the state government of Virginia in 1934, Joseph DeJarnette, a leading member of the Virginia eugenics and sterilization movement, argued that the state needed to extend the sterilization law to more closely resemble the comprehensive German law.[38]

Concern about the racial degeneration of the American population was also expressed at the annual meeting of the New York State Association of Elementary School Principals in 1934, which called for increased sterilization of "criminals and low mentality classes." A 1937 survey by *Fortune Magazine* indicated that a majority of the American population supported the extension of sterilization. Sixty-six percent were reported as supporting compulsory sterilization of habitual criminals.[39] Marion S. Norton, the leading figure of the Sterilization League of New Jersey, defended the Nazi law against attacks by the American Catholic Church and declared it a model for the United States.[40]

Leon F. Whitney, secretary of the American Eugenics Society, was similarly supportive of Hitler's race policy. In a note sent to several newspapers in 1933, Whitney, speaking for the American Eugenics Society, claimed that Hitler's sterilization policy had demonstrated the Führer's great courage and statesmanship.[41] Though he harbored doubts about the German government's ability to fully implement the law, he described the measures as evidence that "sterilization and race betterment are . . . becoming compelling ideas among all enlightened nations."[42]

The widespread support of the American eugenics movement for Nazi sterilization regulations is also evident in numerous articles in *Eugenic News*. Edited in the 1930s by Harry Laughlin, Charles Davenport, and Roswell Johnson, it served as the official organ of the three major eugenics societies in the United States. In 1934, it reported that in "no country of the world is eugenics more active as an applied science than in Germany" and praised the Nazi sterilization law:

> One may condemn the Nazi policy generally, but specifically it remained for Germany in 1933 to lead the great nations of the world in the recognition of the biological foundations for national character. It is probable that the sterilization statutes of the several American states and the national sterilization statute of Germany will in legal history, constitute a milestone which marks the control by the most advanced nations of the world of a major aspect of controlling human reproduction, comparable in importance only with the states legal control of marriage.[43]

In an article about the Nazi sterilization law, the journal reported that the legislation provided any individual who considered her- or himself as hereditarily ill with the possibility of applying for sterilization. Petitions would be decided on a case-by-case basis by a special court that represented the "eugenic interest of the family stock of the Reich." The law was to be equally applied to all "hereditary degenerates . . . , regardless of sex, race, or religion.'' *Eugenic News* concluded:

> The new law is clean-cut, direct and a "model." Its standards are social and genetic. Its application is entrusted to specialized courts and procedures. From a legal point of view, nothing more could be desired.[44]

After the beginning of sterilization in Germany on January 1, 1934, *Eugenic News* printed the translation of an article from a German journal about the success of steps taken to implement racial hygiene in Nazi Germany. The author stressed that, with special loans for married couples, higher taxes for single persons, hereditary homestead laws, and the introduction of labor camps, the measures of the Nazi government were "made for the benefit of posterity, regardless of the approval or disapproval of present generations." Especially important was the Law on Preventing Hereditarily Ill Progeny, which, the author argued, had created a "tremendous sensation all over the world. Scientists and laymen . . . greeted it enthusiastically as a milestone in the history of mankind and as return from a hitherto wrong path." The law would ensure that "healthy, happy generations can now live and develop with the protection of the state." The article concluded with a quote from an unnamed American scientist: "Germany has made world history with her sterilization law!"[45]

Until 1939, *Eugenic News* regularly reported on the development of German eugenics policy. It issued a translation of the German sterilization law and informed readers about the number of sterilizations in Germany. It also printed a translation of a speech by Wilhelm Frick, Reich minister of the interior, about Nazi race policy, and it congratulated Freiherr von Verschuer for founding the Institute for Hereditary Biology and Racial Hygiene at the University of Frankfurt. In addition, it reprinted—without commentary—a translation of an article of the *Rassenpolitische Auslandskorrespondenz* claiming that the percentage of Jewish physicians in Berlin was too high.[46]

The *Rassenpolitische Auslandskorrespondenz* was clearly conscious of the importance of *Eugenic News* n 1935, it proudly announced that the extensive and detailed coverage of eugenic laws in Nazi Germany by *Eugenic News* was an "unambiguous" sign of the "highest attention" foreign scientists bestowed on Nazi Germany. In 1936, *Volk und Rasse,* an "illustrative monthly journal for German national tradition, race science, and race care," published by the Reich Ministry of the Interior and the German Society for Racial Hygiene, complained that *Eugenic News* had published inaccurately high figures regarding the number of enforced sterilizations in Germany and expressed its hope that the journal would correct its mistake in the next edition. A correction and an apology did appear in the next edition of *Eugenic News*.[47]

Nazi Sterilization Propaganda in the United States

The American eugenics movement was especially impressed by Nazi propaganda that promoted the ideals of race improvement. Harry Laughlin coordinated efforts to introduce Nazi sterilization propaganda to the American public. An enthusiastic supporter of Nazi Germany, he had collected newspaper clippings about the National Socialists even before their rise to power in 1933. On the margin of a clipping concerning the opening of a Nazi racial bureau for eugenic segregation, the words *Hitler should be made honorary member of the ERA* (Eugenics Research Association) are written.[48] Laughlin also used his position as assistant director of the Eugenics Record Office in Cold Spring Harbor to organize the dissemination of Nazi race propaganda. He was impressed by the modern methods of Nazi race propaganda, especially by the use of films as a persuasive medium for propagating eugenic goals.

In 1936, Laughlin purchased an English version of the movie *Erbkrank* [Hereditary Defective], an important sterilization propaganda film of the Racial Political Office of the Nazi Party, in order to show it at the Carnegie Institution in Washington.[49] *Erbkrank,* which the Nazi censor administration evaluated as "national political valuable" [*staatspolitisch wertvoll*], and which received "the particular acknowledgment of the Führer," was the basis of a large propaganda campaign in Germany.[50]

The film is introduced by a quote from Walter Gross, director of the Racial Policy Office:

> A people that builds palaces for the descendents of drunks, criminals, and idiots, and which at the same time houses its workers and farmers in miserable huts, is on the way to rapid self-destruction.[51]

Among other pictures of mentally handicapped people the film shows "four feebleminded siblings" who have cost the state "together during more than eighty years of institutionalization 153,000 marks." One subtitle claims that "many idiots are deep under the animal." The film concludes with the statement "that the prevention of hereditarily sick offspring is a moral duty'' and means ''the highest respect for the God-given natural laws.'' The film ends with a picture of a man and a women planting with the subtitle: ''The farmer, who prevents the overgrowth of the weed, promotes the valuable.''[52]

Laughlin described the movie as confined to the ''problem of hereditary degeneracy in the fields of feeblemindedness, insanity, crime, hereditary disease, and inborn deformity.'' He stated that the film:

[C]ontrasts the squalid living conditions of normal children in certain German city slums with the finer and costly modern custodial institutions built for the care of handicapped persons produced by the socially inadequate and degenerate accompanied by captions explaining the family history and description of the near-kin of the particular subject.

Although the film propagated the notion that Jews were particularly susceptible to mental retardation and moral deviancy, Laughlin asserted in *Eugenic News* that the picture contained ''no racial propaganda of any sort.'' The film's sole purpose, he argued, was to ''educate the people in the matter of soundness of family-stock quality— physical, mental, and spiritual—regardless of race.''[53]

Impressed by the film's powerful effect on the audience at the Carnegie Institution, Laughlin decided to use a slightly altered version of *Erbkrank* to help educate the wider American public about race improvement. His assistant, Alice M. Hellmer, informed the S.S. Office for Race and Settlement about this plan.[54] Laughlin raised money to fund the distribution of the film's edited version, entitled *Eugenics in Germany,* to churches, clubs, colleges, and high schools. Laughlin wrote to the millionaire Wickliffe Draper:

If this film reflects accurately the policy of modern Germany, that nation in this particular field of applied negative eugenics has evidently made substantial progress in its intention to act fundamentally, on a long-term plan, for the prevention, so far as possible, of hereditary degeneracy.[55]

Draper and his Pioneer Fund undertook to finance the distribution of the Nazi movie. In cooperation with the Eugenics Research Association, the Eugenics Record Office sent a flier advertising the film to biology teachers in 3,000 high schools. The Pioneer Fund, the Eugenics Record Office, and the Eugenics Research Association anticipated a favorable response because of the attractive medium and the low cost to viewers.

The movie played twenty-eight times between March 15, 1937,

and December 10, 1938. Although plans for national distribution were never realized, the Nazi press reported that *Erbkrank* had been a success in the United States. An article in a Nazi newspaper, entitled "Rassenpolitische Aufklärung nach deutschem Vorbild: Grosse Beachtung durch die amerikanische Wissenschaft" [Racial Political Propaganda on the German Model Receives Great Attention among American Scientists], reported that the movie had made an "exceptionally strong impression" on American eugenicists.[56]

Reasons for the Support

Why did the eugenics movement in the United States as a whole, ranging from moderate eugenicists to the race anthropologists around Madison Grant, express such enthusiasm for the Nazi sterilization law? In the view of the American eugenicists, the Nazi government had avoided mistakes that were made in the formulation of sterilization laws in the different states of their own nation. The German government had introduced a nationwide, well-conceived law, unlike the heterogeneous, improvised state laws in America.

American eugenicists also viewed the Law on Preventing Hereditarily Ill Progeny as grounded on scientific results. After returning from her study tour of Nazi Germany, Marie E. Kopp claimed:

> [T]he German Law is based on thirty years of research in psychiatric genealogy which was undertaken under the leadership of Dr. Ernst Rüdin at the Kaiser Wilhelm Institute of Psychiatric Research in Munich.[57]

Whitney reported that "from his considerable correspondence with certain German scientists who ever since the war have been enthusiastic advocates of sterilization," he was convinced the law had been based on long years of scientific research. He referred to Germany as a country that knew more about its hereditary "defectives" than any other nation, and pointed to the fact that German research in psychiatry and applied psychology was several years ahead of the United States.[58]

In the *Journal of Heredity*, Popenoe dismissed any charges that National Socialists were racists. He instead stressed the fact that Hitler had formulated his policy after carefully studying the textbook of Baur, Fischer, and Lenz—probably the most popular eugenics textbook in the world. Popenoe believed that the Nazi sterilization law could be seen as a sign that scientific leadership was gaining more and more importance within the Nazi hierarchy.

American eugenicists thought of the German law as legally so well conceived that abuse would be nearly impossible. *Eugenic News* claimed that "to one acquainted with English and American Law, it is difficult to see how the new German Sterilization Law could, as some have suggested, be deflected from its purely eugenic purpose."[59] American eugenicists were impressed by the clear definition of hereditary illness and the polished legal and bureaucratic system surrounding the sterilization law. They referred to the establishment of special Hereditary Health Courts and appellate courts as a means of protecting both individual and societal interests.

Another reason for American support of the Nazi sterilization law was based on the fact that the German law distinguished eugenic sterilization from the use of sterilization as punishment for criminals. Nevertheless, castration was accepted among eugenicists in both countries when this procedure was clearly distinguished from eugenic-motivated sterilization. The German Law against Dangerous Habitual Criminals, passed November 24, 1933, was welcomed as a contribution to the "battle against dangerous habitual criminals." The law was applied, according to *Eugenic News* in 1934, "to such habitual criminals who would rather break the law than live by honest work, thus making crime a profession or continuous source of income." The law sanctioned castration for "dangerous sex offenders" with at least two perpetrations of certain sex offenses. Stressing that the law had been designed after the pattern of model foreign laws, *Eugenic News* expected "that the energetic and moderate handling of this law will effectively help the fight against habitual and professional crime."[60]

German propaganda used the medical, legal, and bureaucratic elaboration of the German sterilization law to contrast the Nazi measures with the chaotic situation of sterilization in the United States. Up to the early 1930s, German propaganda referred to abuse of sterilization in the United States. Such propaganda stressed the exemplary, legally correct implementation of sterilization in Germany by focusing on the arbitrary sterilization of inmates in American mental hospitals and jails.[61]

The main reason why the Law on Preventing Hereditarily Ill Progeny gained nearly unanimous support among American eugenicists was due to the fact that the law did not sanction sterilization based on ethnic or religious background.[62] Leon F. Whitney assumed that "American Jewry is naturally suspecting that the German chancellor had the law enacted for the specific purpose of sterilizing the German Jews." He claimed that the German law provided for sterilization of hereditary defectives only, and that it stated that "the measure is solely eugenic in its purpose, and were it not for its compulsory character, it would probably meet with the approval of all who are free from religious bias."[63]

American eugenicist Robert Cook also forcefully denied rumors that the sterilization law was used as a "racial purifier" against Jews and other ethnic minorities. He accused opponents of sterilization measures in the United States of using resentment against the Nazis to "kill two birds with one stone." He pleaded for an "objective view" on the sterilization law, which would preclude any possibility that the law would be abused due to race, class, or gender biases. Heinrich Schade, a fervent supporter of Nazi race policies at the Institute for Human Heredity and Racial Hygiene in Frankfurt, reviewed Cook's article and found it "very objective and accurate." He applauded the fact that Cook countered misleading "horror stories" that depicted the sterilization law as a "racial purifier."[64]

The consensus among eugenicists was so strong that Richard Goldschmidt, a liberal Jewish eugenicist who was forced by the Nazis to flee his position at the Kaiser Wilhelm Institute for Anthropology, Human Heredity, and Eugenics, complained that the Nazis "took over our entire plan of eugenic measures."[65]

American Eugenicists
in Nazi Germany

> Without attempting to appraise the highly controversial racial doc-
> trine, it is fair to say that Nazi Germany's eugenic program is the most
> ambitious and far-reaching experiment in eugenics ever attempted by
> any nation.[1]
>
> Lothrop Stoddard after a visit
> to Germany in 1940

American eugenicists followed the development of eugenics measures
in Germany with rapt attention. The widespread coverage of the Nazi
eugenics program by the American mass media promised to promote
eugenics within the United States, as long as the eugenics movement
succeeded in separating the positively regarded eugenics measures in
Germany from the negative image of the barbaric, racist Nazis.

The secretary of the American Eugenics Society, Leon F.
Whitney, boasted that eugenics had enjoyed a steady increase in public
interest since 1934. He explained the increase primarily as the result of
in-depth coverage by American newspapers of Hitler's plan to sterilize
400,000 Germans, about 1 percent of the population. He viewed
Hitler's extreme measures as creating discussion "among thousands of
persons [in the United States] who may never before have taken any
real interest in the subject."[2]

For American eugenicists, it was important to acquire firsthand
information regarding the application of measures within Germany. In
their view, the success of German eugenic laws resulted from the
consistent application of an exemplary legislative design. In order to
promote Nazi race policies, American eugenicists attempted to collect
information and evidence with which they could counter criticism,
such as the rumors that the purpose of the German sterilization law was
to eliminate the Jewish population.

After the National Socialists seized power in 1933, American eugenicists frequently traveled to Germany. Nazi race and health administrators offered American eugenicists, anthropologists, psychologists, psychiatrists, and geneticists who visited Germany the "proof" they were seeking to convince themselves of the scientific and political merits of Nazi race policies. The administrators facilitated access to high-level Nazi politicians and scientists, and arranged for American scientists to visit institutes, public health departments, and Hereditary Health Courts. German racial hygienists were instructed to treat their foreign visitors with special courtesy. We can only estimate the extent to which American eugenicists increased their professional contacts with German racial hygienists after 1933. The generally enthusiastic coverage, however, given to visits of American eugenicists by German newspapers reveals the importance of such contacts for Nazi Germany.

American Eugenicists Visiting Early Nazi Germany

The first American eugenicist who came to Germany to witness the application of eugenics measures was William W. Peter, secretary of the American Public Health Association. In 1933 and 1934, he traveled through Germany for six months on a stipend from the Karl Schurz Foundation. During his journey, he visited nearly every major section of Germany and met with officials who were responsible for the new health and race measures. He also participated in conferences at which German physicians were trained to assume their new duties as "Rassenpfleger" [racial purifiers] of the German people. Peter's experiences convinced him that the Nazi government was applying eugenics measures in a "legally and scientifically fair way." He judged Germany as the "first modern nation to have reached a goal toward which other nations are just looking, or approaching at a snail's pace."[3]

Back in the United States, Peter deposited a version of information and impressions he recorded with the American Public Health Association and published an article in the *American Journal of Public Health and The Nation's Health*. In this article he reassured the American scientific community that several safeguards would prevent potential abuse of the Law on Preventing Hereditarily Ill Progeny. The Hereditary Health Courts and Hereditary Health Supreme Courts, he argued, would guarantee the correct application of the law. Peter was convinced that the system guaranteed a fair trial for "unfit" members of German society.

The reasons for the sterilization program were readily apparent to Peter:

Germans must live with themselves within their own borders for the immediate future, and depend more than ever upon their own resources. These resources are much depleted. Hence the present load of social irresponsibles are liabilities which represent a great deal of waste.

Economic problems, including reparations, public and private postwar debts, the loss of colonies, declining foreign trade, the painfully slow recovery from inflation and unemployment, and senior citizen pensions, necessitated the sterilization of the majority of the nation's 700,000 handicapped people. Peter explained to his readers that "to one who lives [in Germany] for some time, such a sterilization program is a logical thing."[4]

The tone of Peter's article, written for the American scientific community, contrasted sharply with an article he published in *Neues Volk*, a popular journal of the Racial Political Office of the National Socialist Party. Peter told his German readers of a dream he had about a procession of people, many of them crippled. In that dream the few strong people were forced to carry the crippled, blind, and deaf on their backs. Horrified at this spectacle, Peter asked a strong, blond man why he had to cater to the handicapped. The blond man told him that he did not know the reason, but that he had carried the burden since his youth and would go on carrying it until his death. He explained to Peter that the handicapped were of diverse races and cultures. He complained that they also had many more children than the healthy. The old man on his back, for example, had ten children.

Deeply disturbed, Peter inquired, "Why do you not undertake anything to hinder hereditarily ill people from reproducing more of their kind?" Suddenly, he noticed a group in the procession with a much lighter burden, including hundreds and hundreds of people who had nothing at all on their backs. He saw nearly no handicapped babies or small children. He asked the first man he encountered how they managed to have so few handicapped children. The man answered that, since 1934, they had curtailed the reproduction of all the "misfortunate." He explained to Peter how they had constructed social institutes staffed by doctors and lawyers who decided each sterilization case individually. He asked Peter if he could imagine anyone who would like to raise a child who would be nothing but a burden throughout his or her life.

Peter explained to his readers that, although the story was only a dream, one question haunted him: "How long, how long shall this huge burden of misery press down upon mankind?" He concluded that "an overtaxed world waits hopefully for the result of the recent great enterprise in Germany, which shall enrich human life."[5]

The person who studied German race policy in the greatest detail and who did a lot to propagate the advantages of Nazi race politics within the American eugenics movement was Marie E. Kopp. Kopp, who was not very well known in the scientific world before her comprehensive publications about Nazi race policy, had visited Germany for six months in 1935 on an Oberländer Fellowship. Her job was to conduct a detailed study of the origin, application, and impact of the German eugenics policy and to compare it to what she had observed on a prior journey in 1932. Through contact with scientists at the Kaiser Wilhelm Institutes, she received permission in 1935 to interview superintendents of hospitals and other health institutions, judges of the Hereditary Health Courts, and a large number of physicians, surgeons, psychiatrists, and social workers. She was allowed to visit any public service agency related to her study.

After her return to the United States, Kopp influenced the debate about eugenic measures by publishing speeches and articles in major eugenic, medical, sociological, and criminological journals. She stressed the importance of Germany as the "first country in the world to put an extensive eugenics program into operation among its 65 million people." Such racial measures "were imperative to correct conditions undermining the health of the nation."[6]

In a presentation at the Twenty-fifth Meeting of the Eugenics Research Association in which she compared the sterilization laws of several countries, she stressed that one of the unique features of the German law was that it was not limited to persons in institutions. Only in Germany would the "outward manifestation of a hereditary disease bring every individual under provisions of the law, irrespective of class, race, and creed."[7] At another meeting, she reported to members of the eugenics and sterilization movement that:

> Sterilization law is accepted as beneficial legislation, designated to minimize the difficulties of the afflicted. All possible safeguards are taken to forestall miscarriages of justice in whatever form they may occur. . . . I am convinced that the law is administered in entire fairness and that discrimination of class, race, creed, political, or religious belief does not enter into the matter. I say this with confidence because I had the rare opportunity to examine case histories in large number in various sections of the country and to familiarize myself with the proceedings of the Hereditary Health Courts.[8]

Kopp reported that medical circles in Germany and elsewhere believed that properly performed sterilization operations had no adverse effects on health. She was especially impressed that only 0.4 percent of all women died during sterilization operations. Kopp explained this "remarkably low rate," a total of 4,500 women, by pointing to the fact that most of those sterilized were in good physical health.[9]

Kopp considered the Sterilization Law as part of a comprehensive eugenics program in Germany. She regarded "positive" eugenic measures—the encouragement of racially pure, healthier, superior human beings—as especially important. She explained that the 1933 Law to Reduce Unemployment reduced "male unemployment by relegating back to the home the underpaid women workers." The reemployment of men would encourage them to marry earlier while giving women at home the possibility to raise the children under proper conditions. Kopp explained that marriage loans were given to healthy men when their wives relinquished employment and agreed not to engage in paid work until after full repayment of the loan. Loans of up to $480 could be attained by marrying. With the birth of each baby, the state waived 25 percent of the original loan.[10]

More prominent eugenicists, including two presidents of the American Eugenics Research Association, also traveled to Germany in order to obtain firsthand information about the implementation of Nazi race policies. Clarence Campbell visited racial hygiene institutes while attending the International Population Congress in Berlin in 1935. He became convinced that the conference presentation by the National Socialists had accurately described the theoretical goals and demonstrated how theory was translated into the practical reality of race improvement.

Campbell's successor as president of the Eugenics Research Association, banker and millionaire Charles M. Goethe, was equally enthusiastic about the development of the race policies in Nazi Germany. Goethe was a founder of the Eugenics Society of Northern California and the Immigration Study Commission, which lobbied to extend the Johnson Immigration Restriction Act to include Latin America. He was especially interested in restricting Mexican immigration. Goethe, who worked closely with Popenoe and Gosney, was a trustee of the Human Betterment Foundation, a member of the advisory board of the Sacramento Mental Health Association, and a member of the advisory board of the American Genetics Association. He funded eugenics organizations on both the East and West coasts. As president of various firms and of the Goethe Bank, Goethe traveled frequently to Germany. He used his annual business trips to study the progress of eugenics measures. Goethe reported his findings to the Eugenics Research Association and Human Betterment Foundation and called for a similar policy to be enacted in the United States.[11] He wrote to Gosney about how the United States and Gosney personally had contributed to eugenic developments in Germany:

> You will be interested to know that your work has played a powerful part in shaping the opinions of the group of intellectuals who are behind Hitler in this epoch-making program. Everywhere I sensed that their opinions have been tremendously stimulated by American thought, and particularly by the work of the Human Betterment Foundation. I want you, my dear friend, to carry this thought with you for the rest of your life, that you have really jolted into action a great government of 60,000,000 people.[12]

Another important figure of the eugenics movement, Marion S. Norton, visited Germany in 1938. She studied sterilization laws and efforts to legislate sterilization in Switzerland, Denmark, Sweden, Finland, and England. Norton became well known in Germany for pamphlets she published and for her role as a major figure of the Sterilization League of New Jersey. The tone of a pamphlet that she published in 1937, entitled *Selective Sterilization in Primer Form,* reveals the similarities between the sterilization propaganda of her own organization and the racial propaganda of National Socialism. One of the photographs in the pamphlets carried the following caption:

> See the happy moron;
> He doesn't have a care,
> His children and his problems
> Are all for us to bear.[13]

In 1935 Norton published a small pamphlet entitled *Sterilization and the Organized Opposition,* an angry counterattack against Catholic critics of sterilization. The pamphlet received special attention among German racial hygienists and politicians. She countered claims that sterilization in Germany was used to eliminate Jews, and complained that such rumors stemmed from the ''cunning effort on the part of Catholics to emotionally stampede people, who otherwise would support a measure for social health.'' She criticized the ''strangulating power'' of Pope Pius XI's opposition to all forms of population control that he reiterated in his Encyclical *Casti Connubii* on December 31, 1930. Nazi race politicians used Norton's pamphlet when confronting Catholic opposition to the Law on Preventing Hereditarily Ill Progeny. On September 18, 1937, Gottfried Frey, member of the health department of the Reich minister of the interior, wrote a report for the Reich Chancellory about the "fight of the Catholic Church against the Law on Preventing Hereditarily Ill Progeny." In the report he quoted Norton to prove that the opposition of the Catholic Church was coordinated "from abroad (Rome)" rather than from Germany. Frey also argued that Norton had unveiled a change in the strategy of the Catholic Church. After first attacking sterilization laws as incompatible with moral and religious standards, the Catholic hierarchy had switched to attacking the scientific basis of eugenics measures.[14]

Because of her renown, Norton was cordially received by Falk Ruttke. She gladly fulfilled his request to compose English captions for a German sterilization film called *The Fatal Chain of Hereditary Disease*. Impressed by the movie, she purchased a copy for the Sterilization League of New Jersey. In addition, she ordered twenty-seven photographs used by German race propagandists. Convinced that the German people saw sterilization as an individual "benefit, not as punishment," she continued to promote Nazi race policies abroad. After returning to the United States in the summer of 1938, she published her experiences in an eight-page pamphlet entitled *Major Provisions for Population Control Abroad*. This pamphlet included summaries of the Law on Preventing Hereditarily Ill Progeny and other sterilization laws.[15]

American Eugenicists in Germany during World War II

Even after the beginning of World War II, American eugenicists continued to visit Germany. In the winter of 1939–1940, for example, American geneticist T. U. H. Ellinger visited Germany, apparently undeterred by recent displays of Nazi aggression. He met with Hans Nachtsheim, a geneticist at the Kaiser Wilhelm Institute for Anthropology, Human Heredity, and Eugenics, and examined Nachtheim's elaborate experiments on the genetic causes of disease. During a meeting with the director of the Institute, Eugen Fischer, Ellinger was also introduced to Wolfgang Abel, who wore the black uniform of the S.S. Abel was an anthropologist known for his research on Gypsies and bushmen and had been active in sterilization campaigns directed toward blacks. He later served as an adviser to the S.S. concerning the problem of "Germanizing" Nordic elements in the Russian population. Abel presented Ellinger with detailed information about research on the "Jewish element" in the German population.[16]

After his return to the United States, Ellinger explained to readers of the *Journal of Heredity* that the treatment of Jews in Germany had nothing to do with religious persecution. Rather, it was entirely "a large-scale breeding project, with the purpose of eliminating from the nation the hereditary attributes of the Semitic race." He compared the relationship of German science to the brutal treatment of Jews with that of American scientists and the black population. He viewed the decision to eliminate inferior "hereditary attributes" from the nation and the decision to support the procreation of "Nordic elements" as strictly a matter of politics that had nothing to do with science. "But," he concluded, "when the problem arises as to how the breeding project may be carried out most effectively, after the politicians have decided upon its desirability, biological science can assist even the Nazis."

Ellinger was impressed by the "amazing amount of unbiased information" collected by the Kaiser Wilhelm Institute regarding the physical and psychological defectiveness of Jews. Ellinger believed that the Institute could play an important role in defining whether or not a person had Jewish ancestors. He wrote:

> If it be decided by the Nazi politicians that persons with Jewish ancestors shall be prevented from mating with those who have not such ancestors, science can undoubtedly assist them in carrying out a reasonably correct labeling of every doubtful individual. The rest remains in the cruel hands of the S.S., the S.A., and the Gestapo.

Ellinger speculated that the idea behind the cruel treatment of the Jews "might be to discourage them from giving birth to children doomed to a life of horrors." In 1942, the year that witnessed the installation of gas chambers in Auschwitz, Ellinger argued that if the cruelties "were accomplished, the Jewish problem would solve itself in a generation, but it would have been a great deal more merciful to kill the unfortunate outright."[17]

Characteristically, the only critic of Ellinger's article in the *Journal of Heredity* was a Jewish emigrant from Germany, Richard Goldschmidt, who had been a former assistant director of the Kaiser Wilhelm Institute for Anthropology, Human Heredity, and Eugenics. Goldschmidt criticized Ellinger's "naïve paean of praise" for the Nazis. Based on his own experiences and his dismissal from the Institute, Goldschmidt argued that the Nazis prostituted science. Furthermore, distinguishing between Germans and Jews was impossible, he argued, because Germans as well as Jews encompassed a mixture of genetic elements.[18] Goldschmidt, however, who also maintained that Nazism was better than Bolshevism,[19] did not challenge the basic assumptions of Ellinger's article: the belief in the need for race improvement.

At the same time that Ellinger was studying Nazi race policies, Lothrop Stoddard also spent four months in Germany. Nazi propaganda pointed proudly to the fact that such famous American eugenicists were visiting Germany, even after the outbreak of war. Stoddard traveled officially as a journalist for the North American Newspaper Alliance, but he used his reputation as a famous eugenicist and racial anthropologist to gain access to the highest ranks of the Nazi hierarchy. Stoddard's books, particularly *The Rising Tide of Color Against White-World-Supremacy* and *The Revolt Against Civilization,* won him renown throughout the white world. He was praised, for example, by President Herbert Hoover, and he testified before the House Immigration Committee in 1924.[20]

In contrast to his journalistic colleagues, Stoddard was allowed to speak with high-level German officials, such as Heinrich Himmler, chief of the German Secret Police and of the S.S., Joachim von Ribbentrop, minister of foreign affairs, and Richard Walther Darré, head of agriculture. He even met personally with Adolf Hitler. William L. Shirer, an American colleague who had been in Germany since 1934, complained that the Reich minister for propaganda gave special preference to Stoddard because his writings on racial subjects were "featured in Nazi school textbooks."[21]

Stoddard also met with top Nazi racial hygienists, such as Hans F. K. Günther, Eugen Fischer, and Fritz Lenz, and visited their institutions.[22] He was impressed by the comprehensive character of the Nazi race policy. According to Stoddard, Hitler's race ideology consisted of two very dissimilar aspects:

> The first of these concerns differences between human stocks. Hitler assumes that such differences are vitally important and that "the purity of the racial strain must be preserved." Therefore, logically, crossings between them are an evil. This is the Nazi doctrine best described as *racialism.* The interesting thing is that Hitler does not here stop to labor the point. He takes it for granted as self-evident and passes on to other matters. . . . These concern improvement *within* the racial stock, that

are recognized everywhere as constituting the modern science of *eugenics,* or race-betterment.[23]

Stoddard emphasized that in practice the Third Reich concentrated on the regeneration of the German stock. He viewed racialism, on the other hand, as a "passing phenomenon." Stoddard claimed in 1940 that the "Jews problem" is "already settled in principle and soon to be settled in fact by the physical elimination of the Jews themselves from the Third Reich."[24]

Stoddard explained that when the Nazis came to power, Germany was "biologically" in bad shape. The "best stock" had perished on the battlefields of World War I, and economic depression, mass unemployment, and widespread hopelessness had caused the birthrate to drop so quickly that the nation was no longer reproducing itself. The elements of "highest social value" refused to have children, while the "morons, criminals, and other antisocial elements" reproduced "at a rate of nine times as great as that of the general population."

When the Nazis came to power, argued Stoddard, they started to increase "both the size and the quality of the population." They coupled initiatives designed to encourage "sound" citizens to reproduce with a "drastic curb on the defective elements."[25] Stoddard personally witnessed how the Nazis were "weeding out the worst strains in the Germanic stock in a scientific and truly humanitarian way." On the recommendation of German racial hygienists and because of his contacts with leaders of the Nazi Party and the state, he was able to attend a session of a Hereditary Health Supreme Court. For Stoddard, these courts constituted a sophisticated system for the implementation of the Nazi race policies.

At the Hereditary Health Supreme Court in Charlottenburg, Berlin, he joined two regular Nazi judges, a psychopathologist, and a criminal psychologist. Stoddard reported on four cases that he reviewed in order to illustrate the urgency of sterilizations:

1. An "apelike" man with a receding forehead and flaring nostrils who had a history of homosexuality and was married to a "Jewess" by whom he had three "ne'er-do-well children."
2. An obvious manic-depressive, of whom Stoddard wrote that "there was no doubt that he should be sterilized."
3. An eighteen-year-old deaf-mute girl with several "unfortunate" hereditary factors in her family.

4. A seventeen-year-old mentally retarded girl employed as a helper in an inexpensive restaurant.

Stoddard left that court session convinced that the law was "being administered with strict regard for its provisions and that, if anything, judgments were almost too conservative."[26] He was impressed that such measures to eliminate "inferior elements" were also accompanied by financial support for the procreation of "hereditarily valuable" couples and by a systematic program to propagate a "racial and eugenic consciousness."

The first-hand information that American eugenicists carried home from their visits to Germany shaped the image of German race policy within the United States. Their positive impressions of the Nazi policy influenced the scientific community. They countered the negative reports by Jewish and politically progressive German scientists who had been forced to emigrate and who attempted to inform the American public about the inhuman and destructive character of Nazi policy.

Science and Racism: The Influence of Different Concepts of Race on Attitudes toward Nazi Race Policies

German National Socialism, within its favored racial group, has one of the most eugenic attitudes but its ultraracialism otherwise betrays it into dysgenic practice.[1]

Eugenicist Roswell H. Johnson in 1934

Despite widespread support for Nazi race policies within the American eugenics movement, American historical scholarship has traditionally argued that only a small group of eugenicists supported Nazi race policies, and that this group was increasingly marginalized and discredited within the scientific community. Historians like Mark H. Haller, Kenneth M. Ludmerer, and Daniel J. Kevles have differentiated between two groups within the eugenics movement. The first lent support to Nazi policies and shared similar goals with the Nazis—improvement of the white race through the elimination of "inferior" elements and prevention of miscegenation with other races. The second group, which gained increasing influence in the 1930s, assumed a critical stance toward Nazi race policies.

Three historians—Haller, Ludmerer, and Nils Roll-Hansen—have argued that the distinction between these two groups can be explained by understanding their relationship to science. They have argued that the second group of eugenicists was knowledgeable about the latest developments in genetics and in step with modern scientific thought. They view the group that supported Nazi racial hygienists, on the other hand, as having practiced "pseudoscience" in order to support strongly biased political positions.[2]

It can be argued that the distinction between two groups of eugenicists—one supporting, one opposing, Nazi race policy—is too simplistic. Furthermore, the relationship of eugenicists to science cannot be used as an adequate delineating factor. Rather, both groups of eugenicists included scientists of international stature, as well as those who were more involved in translating scientific positions into publicly accessible positions and promoting political action. In the controversies among eugenicists about Nazi race policy, scientific and political positions were deeply intertwined. To separate eugenicists into groups of "scientists" and "pseudoscientists" denies the complex interaction between science and politics within the various branches of the American eugenics movement.[3]

Broadly speaking, eugenicists perceived themselves as both scientists and social activists. Most believed that there should be a close relationship between their research and its political implementation. The division of labor within the eugenics movement, however, reveals a complex picture. The largest group of activists consisted of biologists, geneticists, psychiatrists, sociologists, zoologists, and physicians, who often held prestigious positions within the scientific community. Five presidents of the American Association for the Advancement of Science (AAAS) served as members of the advisory board of the American Eugenics Society. Furthermore, many eugenicists were highly placed in central positions in related professional scientific organizations, such as the American Genetics Association, the National Academy of Sciences, the American Sociological Association, and the American Social Hygiene Association.[4]

A smaller group of eugenicists did not possess this type of scientific background. Eugenicists such as author Albert Edward Wiggam and sterilization propagandist Marion S. Norton used scientific language to bolster their arguments, but they did not view themselves as professional scientists. Another group, composed of very wealthy members of the eugenics movement, understood their role as supporting the movement's endowment. For example, Mary Harriman, her daughter Mary Rumsey, and Charles Goethe provided the movement with extensive resources. They sat on the advisory board of the American Eugenics Society with other benefactors, such as powerful bankers Frank Babbott and Robert Garrett.[5]

However, despite the varying orientations of members within the eugenics movement, almost all saw the scientific and political aspects of eugenics as closely interrelated. Professional gatherings usually addressed practical policy as well as new research. The distinction between "pure" and "applied" eugenics was primarily analytical.

The Artificial Separation between Science and Politics

Some eugenicists tried to distinguish themselves from the more political branch of eugenics. Biologists Ross Harrison, Herbert S. Jennings, Clarence McClung, and Raymond Pearl distanced themselves from the organized movement in the 1920s because they objected to its use of propaganda; however, they continued to adhere to a eugenic ideology. Eugenicists such as Charles B. Davenport, Lewellys F. Baker, a physician from Johns Hopkins University, and Henry Edward Crampton, a zoologist and experimental biologist from Columbia University, pursued a different strategy. They served as key figures of the organized movement and privately supported its political demands. However, they carefully limited their public remarks about the political implications of their work. A case study of two individuals, Davenport and Pearl, who saw themselves as engaged solely in "pure" eugenics, reveals specific patterns of behavior toward Nazi racial hygienists.

Raymond Pearl, professor of biometry and vital statistics at Johns Hopkins University, distanced himself from the organized eugenics movement in an angry article published in *The American Mercury* in 1927. He lambasted the "biology of superiority" taught by many of his colleagues. He asserted that eugenics had:

> largely become a mingled mess of ill-grounded and uncritical sociology, economics, anthropology, and politics, full of emotional appeals to class and race prejudices, solemnly put forth as science, and unfortunately accepted as such by the general public.[6]

Pearl shifted his interest to population science, a field in which he had already done research. He maintained only loose ties with the organized eugenics movement, but clearly did not want to disassociate himself completely from eugenics. In 1935, when William K. Gregory threatened to sever his connection to the eugenics movement if *Eugenic News* did not cease publishing favorable articles on Nazi eugenics, Pearl intervened. Gregory, a paleontologist at the American Museum of Natural History and president of the New York Academy of Medicine, served on the board and advisory council of the American Eugenics Society from 1923 to 1935 and was a member of the executive committee of the Galton Society. In response to his threat to resign from the Galton Society, Pearl urged him to reconsider his decision.[7]

The reasons for Pearl's intervention in this matter are unclear. Although he saw some justification for anti-Semitic policies, he criticized the belief in Nordic superiority.[8] When anthropologist Franz Boas started a campaign in 1935 against the Nazi race policy and asked prominent anthropologists, geneticists, and population scientists to sign a petition against the Nazi ideology of Nordic superiority, Pearl questioned the wisdom of the action Boas proposed:

> I have a strong aversion to round-robins by scientific men, and most particularly where the pronouncement is really, however camouflaged, about political questions or angles of political questions which have more or less relation to purely scientific matters. In my observation such round-robins never do any good in correcting an evil they are supposed or intended to correct, and, furthermore, in my observation they always do harm to the scientific men who sign them and through these men to science itself.[9]

Pearl feared that the close connection of German racial hygienists, geneticists, and population scientists to Nazi politics would damage the reputation of eugenics and population science on an international level. By differentiating between "legitimate" science and politics, he could criticize aspects of the German race policy while continuing to support Nazi science until the late 1930s. He published in the *Zeitschrift für Rassenkunde,* a scientific journal for racial hygiene in Nazi Germany, as late as 1937.[10]

Davenport, director of the Station for Experimental Evolution (1904–1934) and the Eugenics Record Office (1910–1934), made a similar distinction between politics and science in Nazi Germany.[11] Although Davenport campaigned for sterilization laws and for the Immigration Restriction Act of 1924, he generally tried to distance himself from eugenics propaganda. He argued that the eugenics movement should be careful not to connect itself too closely with advocates among eugenicists who favored health and food fads, "fitter family" contests, and baby shows, as well as with proponents of more serious causes like birth control, prohibition, and legislation against child labor.[12]

Davenport's sharp distinction between science and politics allowed him to cultivate and extend his relations with Nazi racial hygienists until 1940, a time by which most other eugenicists had ceased extensive involvement with their colleagues in Germany. Between 1933 and 1940, Davenport published several articles in German journals for racial hygiene. In 1934, he participated in the *festschrift* for Eugen Fischer. As late as 1939 he wrote a contribution to the *festschrift* for Otto Reche, an anthropologist who later became a leading figure in planning the "removal" of "inferior" populations in eastern Germany.[13]

In addition to his publications in German journals, Davenport held positions on the editorial boards of two influential German journals, the *Zeitschrift für Rassenkunde und ihrer Nachbargebiete* and the *Zeitschrift für menschliche Vererbungs- und Konstitutionslehre,* both of which were founded in 1935. His position as editor of the latter allows insight into Davenport's relationship to science in Nazi Germany. When asked by Günther Just, professor of human heredity and eugenics in Greifswald, to join the editorial board of the forthcoming journal, Davenport cabled a one-word reply in May 1934: "Yes."[14] On June 26, 1934, Davenport received a letter from Julius Bauer of the medical department of the University of Vienna. Bauer had originally been considered for the position as second editor. He informed Davenport that he had been prevented from becoming an editor of the journal because scientists of Jewish ancestry were now forbidden to edit publications that appeared in Germany. He urged Davenport to recognize that scientific journals in Germany were strongly influenced by political decisions.[15] Davenport did not respond to Bauer, and his name appeared on the editorial board list of the *Zeitschrift für menschliche Vererbungs- und Konstitutionslehre* until 1939.

In October 1935, Walter Landauer, a German socialist geneticist and eugenicist who emigrated to the United States before 1933, asked Davenport to sign a protest resolution against the politically motivated expulsion of German geneticist Julius Schaxel from the German Society for Genetics. Davenport did not sign the resolution, although he agreed that members of a scientific society should not be expelled for political reasons. He wrote to Landauer that he had "some doubt as to how far the scientific workers in one country should interfere in the public policy of another country even though this policy affects the scientific work of a colleague in that country." He believed that American scientists should help politically persecuted scientists "individually in any way we can, but there is no use to write protests to Mr. Hitler."[16]

Davenport's refusal to acknowledge fully the highly politicized nature of eugenics and genetics in Nazi Germany was also expressed in his refusal to support an initiative to approve a panel devoted to questioning the scientific basis of ethnic racism at the Seventh International Congress for Genetics, planned for 1937 in Moscow. In his answer to representatives of this initiative, he stated that such a discussion would only result in an "angry, political debate" and would "bring genetics into bad repute."[17]

The strict distinction between eugenics as science and eugenics as politics that allowed Pearl and Davenport to cooperate with Nazi racial hygienists was not typical of the eugenics movement as a whole. The majority of eugenicists resisted making distinctions between pure and applied eugenics. Most did not regard the political aspects of their work as problematic because they viewed the practical applications of eugenics as stemming logically and directly from a scientific basis. Because the majority of the eugenicists agreed with the principle of combining biological science with politics, differences in their reactions to Nazi race policy cannot be explained by understanding their specific relationship to scientific research.

Racism as the Core of the American Eugenics Movement

A useful way to distinguish between strands in the eugenics movement is to emphasize their differing conceptions of race improvement. All eugenicists held the idea that it was possible to distinguish between inferior and superior elements of society, but not all traced inferiority directly to an ethnic basis. It can be argued, however, that any attempt to designate a group as inferior, combined with a political agenda of race improvement through discrimination, constitutes racism. This follows a recent trend in scholarship that treats any eugenic discrimination against "inferior" people as an expression of racism.[18]

My understanding of eugenics as closely intertwined with racism relies on a broad conception of racism that extends beyond ethnicity and skin color by distinguishing two forms of racism: ethnic and eugenic. *Ethnic racism* represents "classical racism," the application of hierarchical standards to the taxonomy of human racial groups. Such ideology is based on a "typological," "morphological," or "anthropological" understanding of race. For example, Hans F. K. Günther defined race in a strictly zoological sense, as a "group of human beings, which distinguishes itself through a specific combination of physical and mental traits from every other group of human beings."[19] He distinguished between three races: the black race of the Negroids, the yellow race of the Mongoloids, and the white race of the Europoids. Among whites, the Nordic race was the most superior group. The leading nations of the earth, so argued Günther, are those nations with the strongest hint of Nordic blood.[20]

Eugenic racism is based on a genetic understanding of race. Race in this view is regarded as unity of procreation, preservation, and development. It is an attempt to define group cohesion biologically, but without referring to a fixed typology of qualitative differences.[21] Eugenicists in this tradition focus on the improvement of their own race by eliminating its "negative" traits. Eugenic racism, therefore, is the demarcation of certain elements within a particular race, followed by attempts to reduce these elements through discriminatory policies.[22] Alfred Ploetz, for example, based his theory of the "vital race" on the ideas of Francis Galton, the English founder of eugenics. Ploetz distinguished a vital race from the typological system of races, which, in his opinion, was only a narrow morphological description of species.[23]

Both eugenic and ethnic racism represent attempts to define groups of human beings either by visible differences or through scientific methods, and to distinguish between them physically, psychologically, intellectually, socially, or culturally. These two concepts of race, as historian Gisela Bock has argued, work together: Every "ethnic" or "anthropological" race has a "population genetic" constitution, and every "population genetic" race includes one, several, or all "ethnic" or "anthropological" races.[24] Both concepts of race are therefore based on the belief that human beings are of different value. This more or less arbitrary hierarchical taxonomy is translated into biological terminology and implies that human beings should be treated differently due to their value for their race or society.[25]

The eugenics movements in the United States and in Germany attempted to combine ethnic and eugenic racism into a comprehensive program of race improvement. The main textbooks of both movements, Paul Popenoe and Roswell H. Johnson's *Applied Eugenics* and the German manual, referred to as the "Baur, Fischer, Lenz," expressed the need to improve the race through the elimination of inferior members of "white" or "Nordic" stock and by hindering miscegenation of this stock with inferior races.[26] In the mid-1920s, Hitler summarized his underlying racial ideology in *Mein Kampf* as: Just as "one people is not equal to another," so "one person is not equal to another within one *Volksgemeinschaft* [ethnic community]"; therefore, "the individuals within a *Volksgemeinschaft* must be differently evaluated, especially as regards the right to have children."[27]

Although the combination of ethnic and eugenic racism was facilitated by the common goal of race improvement, the linkage of these two forms of racism in the ideology of mainstream eugenicists in Germany and the United States was not automatic, as Bock has argued.

Rather, ethnic and eugenic racism interacted in complex ways within their specific historical and national contexts in the first half of the twentieth century. We can understand conflicts, dissociations, coalitions, and associations among eugenicists and racial hygienists in part by examining to what extent various factions expressed ethnic racism in their eugenic proposals for race improvement.

Different Concepts of Race Improvement among American Eugenicists

Historians of eugenics have normally differentiated the American eugenics movement into two groups. The first group, called "mainline eugenicists" by Daniel Kevles, favored the elimination of "degenerate" elements within the white race and argued for preventing miscegenation between races. The second group, the so-called reform eugenicists, tried to separate themselves from mainline eugenicists and National Socialists by advocating selection on an individual rather than ethnic basis.[28] This distinction parallels attempts by historians to distinguish between critics and supporters of Nazi race policies; most of the mainline eugenicists have been viewed as supportive, while reform eugenicists have been regarded as critical of Nazi race policies.

Some recent scholarship has recognized the inherently racist character of both groups of eugenicists and has argued that any attempt to distinguish between mainline and reform eugenicists is extremely problematic. Barry Mehler has pointed out that the borders between these two groups were highly fluid, and that it is difficult to firmly situate individual eugenicists in one camp or the other. Furthermore, he argues that the ultimate consequences of the ideas of both groups were often the same.[29]

While agreeing with these points, I also believe that it is essential to retain an appreciation for the differences between various branches of the eugenics movement. Specifically, I see the need to distinguish analytically between three groups within the eugenics movement: mainline eugenicists, racial anthropologists, and reform eugenicists. In addition, a fourth group, which I refer to as socialist eugenicists, had only tenuous connections to the organized eugenics movement in the United States. This group will be discussed in Chapter 7. My differentiation is based on various understandings of race improvement, which, I will argue, shaped the positions of eugenicists toward Nazi race policies. By using these categories, I do not want to imply that the groups had clear-cut group identities. The categories should be regarded as "ideal types" that can help clarify various strands of thought within eugenics, not as entirely distinct, separate entities.

Mainline eugenicists dominated the eugenics movements in the United States, Scandinavia, and Germany until the early 1930s. They believed in white superiority, yet argued that the white race also needed further improvement. Mainline eugenicists explained the inequality between races as the result of superior adaptation by some groups in the struggle for existence. In other words, whites were viewed as more advanced than others in the evolutionary process. Mainline eugenicists, such as Harry H. Laughlin and Leon F. Whitney, thus agreed in principle with the ethnic as well as the eugenic racism implemented in Nazi Germany. Although most mainline eugenicists were anti-Semitic themselves, they were careful not to be too blatant in supporting Nazi discrimination against the Jews. They feared that Nazi anti-Semitism would dominate the perception of eugenics in the United States and would overshadow the more "acceptable" measures, such as sterilization, marriage restrictions for handicapped people, and special support for the procreation of "worthy" couples. Mainline eugenicists sought to redirect public attention to these "exemplary" eugenic measures and tried to minimize the anti-Semitism of Nazi Germany, especially after the passing of the Nuremberg Law against miscegenation in 1935.

In a speech at the conference of the American Eugenics Society in 1937, Maria Kopp tried to shift emphasis away from the anti-Semitic measures of the Nuremberg Laws to what she saw as the more important marriage restrictions placed on the mentally and physically handicapped.[30] Similarly, Laughlin, in an article promoting the previously mentioned propaganda film *Erbkrank,* claimed that "there is no racial propaganda of any sort in the picture," despite a long passage in the film about the connection between Jewishness and a disposition to mental illness.[31]

The group of eugenicists who voiced the strongest support for Nazi Germany was clustered around racial anthropologists Madison Grant, Lothrop Stoddard, and Clarence G. Campbell. American racial anthropologists were closely tied to mainline eugenicists in national eugenics societies. Racial anthropologists, however, based their ideology on the presumption of French race philosopher Arthur Comte de Gobineau that races are innately unequal. Their belief in Nordic superiority was combined with a strong anti-Semitic bias.[32] Racial anthropologists were more explicit than were mainline eugenicists in voicing support both for eugenic racism in Nazi Germany, which concerned all ethnic groups, as well as for racism directed specifically at ethnic and religious minorities. When Grant's second important book, *The Conquest of a Continent,* appeared in 1933, a reviewer for *The New York Times* correctly noted: "Substitute Aryan for Nordic, and a good deal of Mr. Grant's argument would lend itself without much difficulty to the support of some recent pronouncements in Germany."[33] Grant personally arranged for copies of this book to be sent to German eugenicists Eugen Fischer and Fritz Lenz, as well as to Alfred Rosenberg, chief ideologist for the National Socialists.[34] In 1937 the book appeared in a full German translation. In a foreword to the translated version, Eugen Fischer stressed that Grant was "no stranger to German readers of writing on race and eugenics" and added that, in a time when the racial idea has become one of the chief foundations of the National Social State's policies, "no one has as much reason to note the work of this man with the keenest of attention as does a German of today."[35]

Racial anthropologists, however, were conscious of the need to be cautious in propagating the ethnic racism of Nazi Germany too openly. Madison Grant wrote to his friend Laughlin that, although "most people of our type" were in sympathy with Germany's actions, eugenicists had to "proceed cautiously in endorsing them."[36] Grant's letter indicates that the difference between mainline eugenicists and racial anthropologists in their support for National Socialists was primarily an issue of degree and willingness to be outspoken.

Reform eugenicists, grouped around Frederick Osborn, Roswell H. Johnson, and Ellsworth Huntington—all of whom served as presidents of the American Eugenics Society—distanced themselves from the blatant ethnic racism of the racial anthropologists and National Socialists. They argued that biological differences between groups were negligible compared to the much more significant biological differences existing between individuals.

Roswell H. Johnson, member of the board of directors of the American Eugenics Society from 1928 to 1932, laid the ideological basis for reform eugenicists. Focusing his attack on Madison Grant, Johnson criticized the notion that all members of one race were in *principle* superior to those of other races; he labeled such views "ultra-racist."[37] In contrast to racial anthropologists and National Socialists, he developed a concept of "overlap racism." His premise was "that in mental traits some races do differ in a significant degree, although the overlap is so great that individual differences outrank social differences in importance." In other words, although races differed in quality, the high-quality individuals of a lesser race might be superior to low-quality individuals of a higher race. Still, the chances of finding a superior human being were higher among whites than they were among members of other races. Osborn refrained from using the term *overlap racism* to describe the direction he favored for the American Eugenics Society.

As historian Barry Mehler has pointed out, however, the ethnic racism of reform eugenicists was only thinly disguised. Osborn and his colleagues used the "problem" of "differential fertility" as a code for expressing alarm at the supposed decline of white, Northern European stock. Despite the fact that the Indian and Mexican populations were small in the United States, Osborn viewed the high differential growth of Indians and Mexicans as problematic. When eugenicists, who claimed that they preferred individual selection over ethnic selection, lamented the fact that "genetically inferior" populations of the rural South and West were reproducing faster than "hereditarily more valuable" stock in the Northeast, the "genetically inferior" population they discussed was clearly conceived of as predominantly black, Indian, and Mexican.[38]

The racism of reform eugenicists shaped their specific position toward Nazi race policy. Criticism by this group focused on the discrimination against Jews in Germany and on the concept of Nordic superiority. Johnson regretted that the enormous progress of the eugenics movement under Hitler "suffered by being linked with anti-Semitism." He feared that the "excellent eugenic program" adopted by the Nazis would be nullified by the "dysgenic" consequences of discrimination against ethnic minorities. The persecution of hereditarily "superior" Jews would cause Germany to "remain behind the attainment that would otherwise have been theirs."[39]

Osborn also combined criticism of Nazi Nordic arrogance and discrimination against Jews with enthusiasm for Nazi eugenic measures. In 1937, he praised the Nazi eugenic program as the "most important experiment which has ever been tried." Despite his doubts that compulsory sterilization could obtain better results than voluntary sterilization, he called the German sterilization program "apparently an excellent one."[40]

American reform eugenicists' simultaneous criticism of Nazi anti-Semitism and enthusiastic support for the Nazi eugenic program were possible only because of a refusal to recognize the inseparable connection between eugenic and ethnic racism in Nazi Germany. Reform eugenicists stressed that eugenic measures in Nazi Germany needed to be evaluated independently of Nazi totalitarianism and Nazi anti-

Semitism. For example, before the Annual Meeting of the American Eugenics Society in May 1938, Osborn lamented the fact that the public opposed ''the excellent sterilization program in Germany because of its Nazi origin.''[41] Thus, reform eugenicists did not equate the eugenics measures of Nazi Germany with National Socialism, and they believed that Nazi anti-Semitism had nothing to do with the eugenic concept of race improvement. In their minds, the fact that the two issues were linked together in Nazi Germany was merely an unfortunate coincidence.

The Influence of Nazi Race Policies on the Transformation of Eugenics in the United States

> I consider Hitler one of the narrow-minded fanatics who have brought endless misery over this world. The first unrevised edition of his book shows his absolute lack of knowledge. For any intelligent person it is not worthwhile to waste as much as a word in regard to the shallow twaddle about race, since every honest scientist, whose view is not obscured by fanaticism, must be able to see through all the false premises and conclusions.[1]
>
> Anthropologist Franz Boas in 1935

Socialist Eugenicists

The reaction of the eugenics movement to Nazi race policy must be seen within the context of the scientific community in the United States. In the 1930s, important scientific and political groups grew more skeptical about the policy of the eugenics movements toward ethnic minorities. The "scientific basis" for discrimination against blacks and Jews was questioned by prominent figures, such as socialist geneticists Hermann J. Muller and Walter Landauer, liberal geneticist L. C. Dunn, and anthropologist Franz Boas. Muller, Landauer, and other geneticists enjoyed increasing prestige in the scientific community due to the research successes and growing importance of genetics.

In contrast to Boas and Dunn, who were in principle critical of eugenics, Muller and Landauer represented a group of socialist eugenicists who were primarily responsible for coordinating the scientific critique against Nazi race policies. Although socialist eugenicists argued that there were no differences between races, they believed that the human race as a whole should be improved by supporting the procreation of "capable" individuals and preventing the reproduction of "inferior" persons.[2] Socialist eugenicists, therefore, as well as critics like Boas and Dunn, focused their opposition on the ethnic racism of Nazi Germany. They concentrated their attacks on dismantling the scientific basis of Nazi anti-Semitism, the ideology of Nordic superiority, and the Nazi policy of prohibiting miscegenation. They argued that the Nazis abused science for their political purposes. Muller, for example, warned:

> There is not one iota of evidence from genetics for any such conclusions, and it is too bad to have them issued with the apparent stamp of genetic authority. They form just the sort of ground which reactionaries desire, on which to raise a pseudoscientific edifice for the defense of their system of sex, class and race exploitation.[3]

The conflict between critical American scientists and Nazi racial hygienists escalated during the preparation for the Seventh International Congress for Genetics. The Congress was originally planned to be held in 1937 in Moscow. In 1936, thirty American geneticists sent a resolution to the general secretary of the Congress, Russian geneticist Solomon G. Levit. They demanded a special section to discuss differences between human races, to explore the question of whether theories of racial superiority had any scientific basis, and to debate whether eugenics measures could lead to any definite improvements in human society. Leading American geneticists signed the resolution. Even some reform eugenicists signed, including Clarence C. Little, president of the American Eugenics Society between 1928 and 1929, and Robert C. Cook, editor of the *Journal of Heredity*.[4]

The resolution produced considerable controversy in Germany. At a meeting at the Foreign Office in August 1936, the Nazi government decided to initiate a broad boycott of the Congress and, if this failed, to send only a small delegation to Moscow. Shortly thereafter, the Nazi government noted with relief that the Moscow regime had canceled the Congress due to its new policy against genetics.[5] The Congress was then postponed until August 1939 and relocated to Edinburgh. Although the meeting concluded prematurely due to the outbreak of World War II, leading socialist eugenicists and geneticists succeeded in preparing a resolution against Nazi race policy—the so-called *Genetico Manifesto*.[6] The *Manifesto* was prepared and supported primarily by scientists from the United States. It demanded effective birth control and the emancipation of women, stressed the importance of economic and political change, and condemned racism against ethnic minorities. The *Manifesto,* however, still adhered to a eugenic ideology:

A more widespread understanding of biological principles will bring with it the realization that much more than the prevention of genetic deterioration is to be sought and that the raising of the level of the average of the population nearly to that of the highest now existing in isolated individuals, in regard to physical well-being, intelligence and temperamental qualities, is an achievement that would . . . be physically possible within a comparatively small number of generations.[7]

The *Genetico Manifesto* clearly demonstrated that the scientists who opposed Nazi race policies did not do so because of opposition to their eugenic orientation. The *Manifesto* signatories were critical only of the arbitrary definition of different races and the discrimination against ethnic minorities.

The struggle within the international scientific community of geneticists concerning the correct position toward Nazi race policy was not between a liberal group of anti-eugenical "real" scientists and a group of reactionary, racist "pseudoscientists." Rather, it was primarily a struggle between scientists with differing conceptions of race improvement and different positions as to how science, economics, and politics should be used to realize their goals.

The Transformation within the American Eugenics Movement

In the 1920s, mainline eugenicists held prestigious positions as professors in universities and as members of leading research institutes, where they received support from major foundations. Their influence extended into the highest political levels of the state and federal governments. The important role they played in shaping immigration policy, health administration, and sterilization laws indicates the extent of their influence as scientific experts in political decision-making. In the 1930s, mainline eugenicists and racial anthropologists lost a large part of this influence. They even lost their dominance over the primary eugenic organization in the United States, the American Eugenics Society, although they continued to dominate the Eugenics Research Association, Cold Spring Harbor Laboratories, and the Human Betterment Foundation in California.

A number of factors contributed to their demise: the deaths of important figures like Henry Fairfield Osborn and Madison Grant; the retirement of Charles Davenport; public criticism of blatantly anti-Semitic statements of eugenicists like Laughlin; discoveries in genetics that contradicted the scientific basis of mainline eugenics; and the demand for a stronger sociological approach to the problems of modern society.[8] Critics outside the eugenics movement heightened these factors by pointing out connections between mainline eugenics and Nazi racial hygiene. The strong support of mainline eugenicists for Hitler's race policies provided critics with powerful ammunition for illustrating the potential consequences of mainline eugenics ideology.

The most prominent critic of Nazi race policy and the position of mainline eugenicists was Franz Boas, professor of anthropology at Columbia University in New York. Boas' efforts to counteract the "vicious, pseudoscientific activity of so-called scientists who try to prove the close relation between racial descent and mental character" have been well documented in the work of Elazar Barkan.[9] Boas sought to publicly reveal the questionable scientific basis of racism, to initiate resolutions by scientists and scientific organizations, and to undertake new research that would "undermine the belief in race as primary factor in cultural behavior."[10]

After several thwarted attempts to organize scientists against Nazi racism in 1933 and 1934, Boas turned toward the more modest goal of gaining support for a resolution condemning the theory of Nordic superiority. Livingston Farrand, president of Cornell University, and Raymond Pearl both refused to collaborate, so Boas turned to a scientist whose views on racial issues were more ambiguous, Harvard anthropologist Earnest A. Hooton. Hooton was one of the world's leading anthropologists and participated in the American eugenics movement, with a special interest in the biological results of miscegenation and criminal anthropology. Although Hooton rejected the theory of Nordic superiority, he was nevertheless somewhat receptive to the racist notions of his colleagues in the eugenics movement. He spoke at the "Fourth National Conference on Race Betterment," organized by Davenport, and wrote to Grant after reading *Conquest of the Continent* that he had "basic sympathy" for Grant's "opposition to the flooding of this country [United States] with alien scum."[11]

Barkan has pointed out that Hooton's attitudes toward Jews were marked by the same ambiguity as his position on race.[12] In an article designed to combat racism, he described Jewishness as a physically determined entity, illustrating his argument with pictures of Jewish noses. At the same time, he emphasized the intellectual superiority of Jews.[13] Hooton stressed his high opinion of Jews, but hoped that they would "strive to eradicate certain aggressive and other social characteristics" which "account for some of their trouble."[14]

Hooton prepared a statement that granted that it was "conceivable that physical races may differ psychologically, in tastes, temperament, and even in their intellectual qualities," but that "a precise scientific determination of such differences has not yet been achieved" and "no definite relation between any physical criterion of race and mental capacity" had yet been found.[15] Hooton's and Boas' hope that this statement would be supported by the leading physical anthropologists in the United States was not realized. It's not surprising that Raymond Pearl declined outright. C. H. Danforth, professor of anatomy in the medical school of Stanford University, also refused to sign. "The Jewish problem in this country is somewhat the responsibility of influential Jews themselves," he argued, who, "because of their Jewish racial solidarity," were "overanxious to retaliate on Germany through America."[16]

After the 1937 International Population Congress in Paris, which led to an escalation of the conflict between Nazi scientists and their critics, Boas and Hooton started a new initiative to establish an independent committee to carry out research on "the whole field of so-called racial behavior."[17] This committee included well-known anthropologists, geneticists, psychologists, and sociologists, as well as reform eugenicist Frederick Osborn. In a letter to Boas written October 11, 1937, Osborn described the attempts of reform eugenicists in the American Eugenics Society to develop a "sound program that will eliminate all of the old class and race biases of eugenics." He argued that the whole question of group superiority should be dropped, and the focus should be placed on "eugenic selection of birth." In other words, the eugenics movement should direct its effort toward the selection of superior individuals, who could be found in every race.[18] Boas reported to Osborn on the new American committee and invited Osborn to participate. Osborn replied that he would be honored to join, but asked that the committee not emphasize German race policy. He suggested that the committee instead focus on the positive amalgamation of a great number of racial types in the United States.

Boas next prepared a resolution, "Proposal for Studies in the Determination of Population Qualities by Genetic and Environmental Factors." This resolution declared that heredity might play an important role in family lines, but that the importance of heredity for large social or racial groups was scientifically unknown. Because of this lack of knowledge, the resolution called for further studies in the field and warned against drawing premature political conclusions.[19] Although some well-known scientists supported the resolution, it generally did not influence the discussion among American scientists.[20]

Osborn and Hooton agreed with Boas only to the extent that they were prepared to criticize Nordic superiority and the scientific justification of anti-Semitism. Hooton and Osborn continued to maintain contact with both mainline eugenicists and their critics.

Osborn and other reform eugenicists, such as Frank Notestein, from the Milbank Memorial Fund, and Warren S. Thompson, president of the Scripps Foundation, were able to basically gain control of the American Eugenics Society when the position of mainline eugenicists within the Society weakened steadily throughout in the 1930s. Through their comprehensive and uncritical support of Nazi race policies, mainline eugenicists had made their own standing in the United States partly dependent on the reputation of Nazi Germany. As Nazism grew more unpopular with the American public, mainline eugenicists were no longer able to distance themselves from Nazi race policies.

Laughlin, for example, was eventually ousted from influential political and scientific positions. The Carnegie Foundation, the sponsor of the most important institutional base of mainline eugenicists, the Eugenics Record Office, accused the Office of producing political propaganda. A Foundation committee that evaluated the Eugenics Record Office in 1935 recommended that it

> cease from engaging in all forms of propaganda and the urging or sponsoring of programs for social reform or race betterment such as sterilization, birth control, inculcation of race or national consciousness, restriction of immigration, etc.[21]

The committee also demanded that the Eugenics Record Office sever its close ties to *Eugenic News*. Even after it adopted a more restrained strategy, criticism of the Eugenics Record Office continued. Finally, in 1939, the Carnegie Foundation forced Laughlin to retire as assistant director. The Office closed on December 31, 1939.

Historians have tended to interpret the difficulties of mainline eugenicists in the 1930s as a crisis of eugenics as a whole. However, a study concerning sterilization in the United States has proven that sterilization for eugenic purposes increased during the 1930s—the same time in which the institutionalized eugenics movement was in a period of redefinition.[22] The transformation of eugenics should be viewed as a shift in power from eugenicists with strong notions of Nordic superiority and anti-Semitism to socialist eugenicists and the reform wing within the American Eugenics Society. Furthermore, the 1930s witnessed a widespread diffusion of eugenic ideology into other scientific fields, such as population science and psychiatry.

The shifts within the American Eugenics Society in the 1930s made it ripe for a transformation into a more sociologically oriented movement. The changes took place peacefully, without an intense internal power struggle, and with general solidarity among the different wings of the eugenics movement. For example, the growing influence of reform eugenicists within the American Eugenics Society did not result in the total exclusion of mainline eugenicists like Laughlin, sociologist Henry Pratt Fairchild, or Henry Perkins, president of the Society from 1931 to 1934. These men remained models for younger eugenicists. Osborn described Laughlin as "a thoroughly competent man of real ability" and Fairchild as a "moderate" who "works well with others of more technical experience."[23]

In his study of the American Eugenics Society historian Barry Mehler has shown that in the 1930s Osborn and Laughlin worked closely together in running the Eugenics Research Association and in editing *Eugenic News*. Laughlin praised Osborn's work on differential fecundity and wrote him that what really counted was the different birth rates "between fine stocks and races on the one hand and incompetent and degenerate races and stocks on the other."[24] Laughlin and Osborn both helped to set up the Pioneer Fund in 1937 and discussed one of its first projects—cash grants to pilots of the Army Air Corps, who were not reproducing their "superior stock" due to a lack of income. Osborn, however, advised Laughlin against running the project by himself. "It would be like a general, responsible for the strategy of the army, wanting himself to drive one of the tanks in the attack," he wrote. Instead, he proposed that they use their contacts to delegate the project to another person.[25]

The shift away from the eugenic aspects of Nazi race policies began quite late. In the late 1930s, the new leadership of the American Eugenics Society became more and more interested in how the eugenics measures in Sweden, which had developed a comprehensive program to support large hereditarily healthy families, could combine with a sterilization law. The American Eugenics Society was searching for a model that combined democracy with eugenics. Mehler has shown that the shift away from interest and support for Nazi Germany was because of opposition to the totalitarian political regime, not because of the Nazi eugenics program.[26] In Osborn's opinion, a democratic welfare state could better guarantee the procreation of the eugenically "superior" elements of society than a totalitarian state. A positive trend, he argued, would materialize the moment large families were provided with enough support. The state, however, should restrain itself, Osborn argued in 1939, from defining people as "unfit" or "fit," "except in the case of hereditary defectives."[27]

The "friendly takeover" of eugenics societies by reform eugenicists opened the movement to new genetic discoveries, new sociological methods, and the question of overpopulation. However, the core of eugenics ideology—the distinction between superior and inferior genetic groups, combined with the aim of race improvement—remained intact. The definition of "inferior" and "superior" groups changed. In 1937 Osborn formulated the position of the American Eugenics Society:

> It would be unwise for eugenicists to impute superiorities or inferiorities of a biological nature to social classes, to regional groups, or to races as a whole.[28]

Osborn and the other leaders of the American Eugenics Society after 1935 did not entirely renounce the notion that there were differences between races, but they did adopt a reformist outlook that diminished the importance of race differences and argued for selection on an individual basis.

The Reception and Function of American Support in Nazi Germany

> Since January 1, 1934, the law of July 14, 1933, on preventing heredi-
> tarily ill progeny has been enforced. Thereby a years-long controversy
> about the admissibility of sterilization, about the way of executing it
> and the extent of its application has been concluded. In a sense, this
> concerns not only Germany but also Europe and the rest of the
> world. . . . German scholarship is still a model to foreign countries
> because for decades it has done pioneering work. Thus, if we do not
> ourselves jeopardize German scientific research and the reputation of
> German training particularly in the field of medicine, this is going to
> remain so in the future, too, and German influence in general is going
> to grow increasingly along with the political regeneration of the Ger-
> man people.[1]
>
> Arthur Gütt, Ernst Rüdin, Falk Ruttke in the
> official commentary about the German
> sterilization law in 1934

Nazi Incentives to Foreign Scientists

In 1934, one of Hitler's staff members wrote to Leon Whitney of the
American Eugenics Society and asked in the name of the Führer for a
copy of Whitney's recently published book, *The Case for Sterilization*.
Whitney complied immediately, and shortly thereafter received a personal letter of thanks from Adolf Hitler. In his unpublished autobiography, Whitney reported a conversation he had with Madison Grant
about the letter from the Führer. Because he thought that Grant might
be interested in Hitler's letter he showed it to him during their next
meeting. Grant only smiled, reached for a folder on his desk, and gave
Whitney a letter from Hitler to read. In this, Hitler thanked Grant for
writing *The Passing of the Great Race* and said that "the book was his
Bible." Whitney concluded that, following Hitler's actions, one could
believe it.[2]

Hitler's personal correspondence with American eugenicists reveals both the influence that American eugenicists had on the highest figures of the Nazi regime and the crucial importance that National Socialists placed on garnering support for their policies among foreign scientists. The Nazi government consistently relied on the support of scientists to propagate their race policies both at home and abroad.

One effective way of enlisting the aid of the international scientific community was to honor foreign scientists by awarding them honorary degrees from German universities. In 1934, for example, the Johann Wolfgang von Goethe University in Frankfurt offered Henry Fairfield Osborn, uncle of Frederick Osborn, an honorary doctorate of science. Henry Osborn was one of America's most famous paleontologists, president of the American Museum of Natural History in New York for twenty-five years, and founder of the department of biology at Columbia University. As president of the Second International Congress for Eugenics in 1921 and founder of the American Eugenics Society, he was one of the earliest important figures in the American eugenics movement. Honored by Frankfurt University's overture, Osborn traveled to Nazi Germany to accept his degree.[3]

Osborn's honor did not receive much publicity in either the American or the German press. Shortly thereafter, however, a much larger-scale attempt to court international scientists attracted widespread attention. In the mid-1930s, the Nazi government decided to use the 550th anniversary of the University of Heidelberg to celebrate the "new spirit" of academia in Germany. Many representatives of non-German universities were invited, and honorary degrees were given to well-known German and non-German scientists.

On this occasion, two renowned eugenicists from the United States were awarded honorary doctorates—Foster Kennedy, a psychiatrist, and Harry H. Laughlin. Kennedy was well known for his membership in the Euthanasia Society of the United States and for his advocacy of the killing of mentally handicapped persons. In 1939, three years after he received his honorary degree in Nazi Germany, he resigned from the Society because he criticized its policy of favoring "voluntary" euthanasia for people who had at one time been well, but later became ill. In contrast, he favored systematic extermination. He thought that it would be for the general good that "euthanasia be legalized for creatures born defective," whose future would be hopeless in the opinion of a medical board.[4] Kennedy continued to promote euthanasia in the United States, even after the mass killings of mentally handicapped people in Nazi Germany had been revealed to the American public.[5]

In May 1936, the dean of the faculty of medicine at the University of Heidelberg and professor of racial hygiene, Carl Schneider, who later served as a scientific adviser for the extermination of handicapped people in Nazi Germany, officially offered Laughlin an honorary degree as Doctor of Medicine. Laughlin's reply conveyed deep appreciation:

> I stand ready to accept this very high honor. Its bestowal will give me particular gratification, coming as it will from a university deeply rooted in the life history of the German people, and a university which has been both a reservoir and a fountain of learning for more than half a millennium. To me this honor will be doubly valued because it will come from a nation which for many centuries nurtured the human seed-stock which later founded my own country and thus gave basic character to our present lives and institutions.[6]

Laughlin, however, did not travel to Germany to accept the award. The official reason he gave to Schneider was lack of time. An important unofficial reason, however, may have been that the participation of American scientists at the jubilee was sharply criticized by the American public media. In April 1936, *The New York Times* called for a wreath to be laid on the "grave of academic freedom" in Germany and claimed that those who went to the jubilee celebration would become "propaganda tools" for the Nazis.[7] Laughlin was probably afraid that a journey to Germany would weaken his position within the Carnegie Foundation, which was already becoming more critical of the work of the Eugenics Record Office. Despite his wariness of being regarded publicly as allied with the Nazi government, Laughlin was proud of the honorary degree. He received congratulations from several colleagues in the eugenics movement and was acknowledged in both the German and American press.[8]

Following the jubilee, Laughlin wrote to Schneider and thanked him again for the "high honor." He requested that Schneider mail the diploma to him in the United States and concluded his letter with the following remarks:

> I consider the conferring of this high degree upon me not only as a personal honor, but also as evidence of a common understanding of

German and American scientists of the nature of eugenics as research in
and the practical application of those fundamental biological and social
principles which determine the racial endowments and the racial
health—physical, mental and spiritual—of future generations.[9]

Laughlin received the degree from the German consulate in New York
City on December 8, 1936. The diploma lauded him as a "successful
pioneer of practical eugenics and the farseeing representative of racial
policy in America."[10]

The cultivation of non-German eugenicists, geneticists, and anthropologists by the Nazi regime was not based primarily on a desire to
acquire scientific information. Rather, the main purpose was to gener-
ate support for a propaganda strategy designed to quell opposition to
Nazi race policies. On the one hand, this propaganda effort was directed at the German population. The Nazis believed that favorable
statements by well-known scientists from other countries would give
the German people the impression that scientific communities abroad
favored Nazi race policy and viewed it as compatible with scientific
knowledge.[11] On the other hand, Nazis relied on the approval of foreign scientists to further their propaganda efforts outside of Germany.
Statements by non-German scientists were more credible than were
those of German scientists, who were often regarded as mere puppets
of the new regime. Thus, the Racial Policy Office informed foreign
guests at the 1936 Olympic Games in Berlin that Nazi race policy was
based on "the internationally accepted" science of racial hygiene and
referred to eugenics measures in the United States and Scandinavia.
Similarly, in 1935, Walter Gross, head of the Racial Policy Office,
informed the international press and foreign diplomats in Germany that
the scientists who had attended the 1934 Congress of the International
Federation of Eugenic Organization approved of the Nazi race policy.[12]

Foreign Reception of Nazi Race Policies

The international reception of Nazi race policies passed through several
phases, at least in the perception of National Socialists. In looking
back, Walter Gross, who was one of the main propagandists of German
race policy, declared in 1939 that Germany's race policies had gained
more and more international acceptance during the previous years.
After countering initial criticism from non-German countries, the Nazi
government and German scientists together succeeded in quelling

much of the opposition by promoting the notion that Nazi race policies were of a scientific character and politically necessary. In a speech at the *Hochschule für Politik,* Gross claimed that this success was among the biggest achievements of the last six years in Nazi Germany, comparable only to the "unheard successes in the development in political, military, or economic areas."[13]

While Gross' speech must be viewed as part of Nazi propaganda, it nevertheless points to a shift in the reaction of foreign countries toward Nazi race policies. From the perspective of Nazi race politicians, the first two years, 1933–1934, were marked by "the devastating and incredible success" of propaganda from Jewish circles that attempted to discredit the scientific character of their race policy.

Gross viewed the Congress of the IFEO in 1934 as the turning point after which the Nazis succeeded in convincing foreign scientists of the scientific character of their race policies. He explained that at the Congress, German and non-German scientists had held long discussions on the question of whether the Nazis really intended to sterilize only people who were hereditarily "ill," or if they would use the law to get rid of former ministers and representatives of the Weimar Republic.

The Nazi journal, *Neues Volk,* summarized foreign criticism under three different categories. First, foreign countries argued that Germans neglected the importance of education and overstressed the importance of inheritance. Second, they believed that National Socialists overestimated differences between races. Third, they feared that Germany's emphasis on the importance of the folk would result in diminished freedom for individuals. *Neues Volk* optimistically believed that such opposition would be overcome:

> The cooperation of the scientists of all nations, who have begun to express agreement with our standpoint, will lead unavoidably in the near future to an intellectual shift in other countries.[14]

Similar evidence illustrating how Germans perceived a weakening of international resistance toward their race policies surfaced in an internal meeting of the Racial Policy Office of Baden in December 1934. The Office representative told his audience that the Law on Preventing Hereditarily Ill Progeny was initially greeted by a "series of outrageous attacks" from foreign countries; however, "only after one year," he stated, "did the defense propaganda start to work."[15] Beginning in 1935, German propagandists voiced greater satisfaction with the prevailing image of Nazi race policies. In June 1935, the Racial Policy Office of the National Socialist Party stated:

> Of greatest importance is the increasing interest of foreign countries in
> the National Socialist race ideology. . . . Scholarly circles in England,
> the United States, and even in Japan have accepted the race ideology of
> National Socialism in a very positive way and are attempting to apply it
> to their own national conditions. The sterilization law is gaining special
> attention.[16]

Likewise, in 1935, Walter Gross claimed at a reception for foreign
diplomats and press agents that respect for the race, health, and population policies of Adolf Hitler was spreading throughout the world.[17]

In a 1938 review article, the *Völkischer Beobachter* emphasized
that after only a few years, protests from foreign countries against
German race and health policy were clearly more subdued. The *Völkischer Beobachter* explained the change by pointing to the fact that
more and more governments were introducing eugenic programs based
on the experiences of Nazi Germany.[18] It referred to the establishment
of sterilization laws in Norway, Sweden, Finland, Estonia, and Canada. However, these were never carried out to the same extent as they
were in Germany, nor were they part of a comprehensive strategy of
race improvement.

Indeed, Nazi propagandists emphasized that the notion that reproduction of "hereditary inferiors" had to be stopped was widespread
and not limited to totalitarian countries. They stressed that politicians
accepted the idea that the "quality" of their people played a central
role in political conflicts even in parliaments and parties of democratic
countries.[19] As part of a report concerning a racial hygiene exhibition
in the United States, the *Völkischer Beobachter* stated that discussions
about German race policy were becoming less emotional and more
scientific and that scientists in the United States were increasingly
convinced that the laws of the Third Reich were exemplary.[20]

Clearly, the Nazi claims that other countries were influenced by
German race policy should be read critically. They were part of a broad
propaganda strategy designed to convince the German population that
Germany served as a model for other nations. The increasing implementation of eugenic laws in different countries in the 1930s may have
been influenced by the Nazis, or they may have been primarily a
reaction to the worldwide Depression. In either case, the increasing
number of eugenics laws in democratic countries does not explain why,
in the perception of the Nazis, the international criticism seemed to
decline. Three other reasons for the decline must be emphasized.

First, as Germany became stronger and more autonomous after

1936, Nazi politicians became bolder in ignoring outside criticism. In 1934 and 1935, Germany was economically and militarily too weak to dismiss other countries' reactions toward its policies. In 1939, Gross recalled the earlier period as an unhappy time when the Reich lacked freedom, power, and economic strength. Gross further explained that only after Germany had attained a strong position in the world could it act freely in the development and propagation of its race policies. Although scientific critics of Nazi ethnic racism and anti-Semitism became better organized and more vocal after 1936, the Nazis were well entrenched by then, and they denounced every critic as a representative of "international Jewry."

Second, criticism was clearly affected by the rising status of Nazi Germany in world affairs. Aware of Germany's increasing influence, some politicians attempted to limit criticism of the Nazis to special fields and to particular forms of expression.

Finally, an important factor for understanding the seemingly diminished and ineffective international opposition resides in the fact that criticism was limited largely to Nazi ethnic racism. The majority of international scientific critics failed to question the aim of race improvement in principle. The disadvantage of this strategy can be seen in the response toward the Law on Preventing Hereditarily Ill Progeny. International scientific criticism focused on the possibility that the law could be abused if applied specifically to ethnic minorities and political enemies. The critics seldom questioned the law as a whole. As they became increasingly convinced that the law was being applied legally, their criticism of this aspect of Nazi race policies tapered off.

Questionable Nazi Distinctions

In viewing American reactions to their race policies, Nazis artificially distinguished between the affirmative reactions of "reasonable" American scientists and negative reactions expressed in the supposedly Jewish-controlled public media. Looking back in 1939, Gross recalled that the leadership recognized a "strange rift" abroad in the early years of the National Socialist regime. While the public media had nothing positive to say about Nazi race policies and criticized the sterilization law, scientific and political representatives were traveling privately to Germany in order to study the results of their policy.

Nazi propaganda explained the negative perception abroad as a result of the "lie campaign" of the "Jewish press." Immediately following the implementation of race policies, "international Jewry" succeeded in denouncing Germany "once again" as the "classical country of barbarism."[21] Nazi propaganda claimed that foreign scientists only gradually succeed in convincing an uninformed public about the beneficial character of Nazi race policies. As Gross recognized, foreign scientists were crucial for changing public opinion abroad. "Their word," he claimed, "was able to balance the opinion of a hundred chatterers."[22]

The Nazis portrayed the United States as a country with an enlightened group of scientists who were thwarted by a Jewish-manipulated public and government. The country with the strongest group of eugenicists, they argued, faced opposition from "international Jewry," which wielded control over large parts of the mass media and government.[23] This image of the United States as divided between "enlightened" scholars and a "manipulated" public helped Nazi propaganda to counter American criticism. In doing so, they were aided by American eugenicists who shared their view.[24]

International Support

In order to influence the scientific community and the public in foreign countries, the Nazi government first needed to mobilize its own German scientists, especially racial hygienists, eugenicists, anthropologists, psychiatrists, and population scientists. However, the relationship between German scientists and Nazi race policy cannot be understood adequately by assuming that the health and race administration of the Nazi state was monolithic or homogeneous. Rather, this totalitarian interpretation should be replaced by an analysis of the Nazi regime as consisting of multiple centers of power. Such an approach allows insight into both cooperation and rivalry among different interest groups and power blocs within Nazi Germany.[25] One can locate several different power groups: Groups clustered around Hitler's chief ideologist Alfred Rosenberg, the Party's Racial Policy Office headed by Gross, the National Socialist Welfare League, and the Nazi Doctors' League under Gerhard Wagner, as well as under the more technocratic and elitist segment of public health administrators, medical researchers, and the S.S.[26]

The relationship between different power blocs in the health and race bureaucracy and racial hygienists should also not be viewed as a simple process of *Gleichschaltung* [coordination] or instrumentalization. The Nazi takeover in 1933 translated into both new possibilities and new risks for administrators, German racial hygienists, anthropologists, and population scientists. Scientists willing to work with the new regime had to negotiate with Nazi leaders. The Nazi demand for total obedience threatened the relatively independent position scientists had enjoyed in the Weimar Republic. However, if racial hygienists could succeed in convincing the National Socialist government of their loyalty, then they could also gain access to new resources and enjoy increased influence as members of prestigious research institutes.

The example of Eugen Fischer demonstrates how individual racial hygienists maneuvered within Nazi ranks. When Fischer's position as director of the Kaiser Wilhelm Institute was threatened as a result of the controversial views of two of his main staff members, Richard Goldschmidt and Hermann Muckermann, Fischer used his prestige in the international scientific community to safeguard his position. In addition to quickly separating himself from Goldschmidt, a Jew, and Muckermann, a Catholic eugenicist, he emphasized his role in international science in a written declaration of loyalty:

> I can truly say that I have gone with all my power to any lengths to serve the most important part of the National Socialistic ideology and politics (human heredity, racial hygiene, population policy). . . . Today the whole world is dealing with the way in which the race question is treated in Germany. . . . Two persons are well known as exponents of the German racial science in foreign countries: Günther and I. Günther is mainly seen as a propagandist. I am called by Schemann "the dean" of this science. I know that in all foreign countries people are following what I said in the past and what I am saying today about race and the importance of the Nordic race. I know what would be thought if I were not allowed henceforth to say anything in public. I would also never say anything against my inner convictions.

The Reich Ministry of the Interior reported to Richard Walther Darré, Fischer's main opponent, that while the Ministry still had reservations about Fischer, it was necessary to retain him. Fischer was an accepted important figure in the field of human heredity and race research both at home and abroad. The Nazi government decided it could not afford to alienate Fischer because a disagreement between him and the administration would give the impression, in Germany as well as abroad, that Fischer had objected to the course of Nazi race policy and that the adopted measures stood in contradiction to scientific knowledge.[27]

The strategy of the German Racial Hygiene Society was similar to that of Fischer and his Kaiser Wilhelm Institute. In 1933, the German Racial Hygiene Society came under the control of the Reich Ministry of the Interior. The minister of the interior, Wilhelm Frick, appointed Ernst Rüdin as the president of the Society:

> I hereby tender you assignment as my honorary representative at the Society for Race Hygiene and the German League for Heredity Science and Racial Improvement. The theory of inheritance and race hygiene are of the utmost importance for the structure of the Reich and for improvement of the race of the German people; therefore, I would like you to carry through the reconstruction work in closest collaboration with my ministry.[28]

After 1933, principal aims of racial hygienists—propagating race improvement among the German people and serving as scientific experts to the government—were either overtaken by Nazi organizations or institutionalized through advisory boards within the different ministries. The racial hygienists supported the Nazi mass organization, joined the advisory boards, and sat in the sterilization courts. However, the only area that racial hygienists could continue to monopolize and dominate was that of promoting German influence within the international eugenics movement. Famous members of the German Society for Racial Hygiene, such as Alfred Ploetz, Rüdin, Fischer, Fritz Lenz, Otmar Freiherr von Verschuer, and Otto Reche, served to boost the international prestige of Nazi racial policies.[29]

Nazi race propaganda portrayed racial hygiene as completely harmonious with the goals of National Socialism. In 1938, Walter Gross claimed that the Nazis had succeeded in establishing close and fruitful cooperation with scientists from the beginning of Third Reich. In no other field, he argued, was cooperation closer than in the fields of population science and racial hygiene, and in no other field was the cooperation between science and politics so imperative:

> In some other fields the scientific community sometimes failed to appreciate our modern times and distracted politicians with outdated notions. However, we can recognize with satisfaction that in the field of race and population policy, German scientists have been loyal collaborators in the implementation of our political aims. . . . The unquestioned unity of politics and science succeeded in promoting German race ideology all over the world, and proved extremely fruitful at several international congresses.[30]

Eugenicists outside Germany could not see the complex character of relations between scientists and National Socialists. They often did

not recognize that power struggles within Nazi Germany were veiled
by an ideology that appeared to show total agreement between Nazi
race policy and science. In one of the first statements of the American
Eugenics Society on the new regime in Germany, Paul Popenoe
stressed that

> Hitler is surrounded by men who at least sympathize with the eugenics
> program. . . . The policy of the present German government is there-
> fore to gather about it the recognized leaders of the eugenics movement,
> and to depend largely on their counsel in framing a policy which will
> direct the destinies of the German people, as Hitler remarks in Mein
> Kampf, ''for the next thousand years.''[31]

The interaction between German and American eugenicists was
characterized by selective perception on both sides of the Atlantic. The
American side associated German eugenic measures with ''scientific,
reasonable'' racial hygienists, separating them from what they saw as
the barbaric racism of only a small ''unscientific'' element in the Nazi
Party. They regarded the close connection between German scientists
and the Nazi administration as a guarantee against possible abuses of
eugenics policies. Likewise, the Germans chose to see outside criticism
of their policies as resulting from a Jewish conspiracy. They refused to
acknowledge that the support, expressed by certain elements within the
eugenics community, was counterbalanced by widespread scientific
opposition to their policies.

The Temporary End
of the Relations between German
and American Eugenicists

A glance at the literature shows that Germany's action [through the
Law on Preventing Hereditarily Ill Progeny] has inspired foreign nations to deal once again with the question of sterilization of hereditarily inferior people. In general, it is possible to say that no other
nation followed the National Socialist law in its rigorous compulsory
character but in principle the problem was accepted as important
everywhere.[1]

Karl Bonhoeffer, arguing for the reestablishment
of the German sterilization law, 1949

The Decline of German–American Relations

When relations between the eugenics movement in the United States
and German racial hygienists began to cool in the late 1930s, it was not
primarily because American eugenicists recognized the negative consequences of the implementation of eugenics principles. Rather, a combination of different factors was at work: gradual recognition by the
public and the scientific community that anti-Semitism was at the core
of Nazi race policy; a power shift inside the scientific community of the
United States toward a group of more progressive socialist eugenicists
and liberal geneticists; and the rapid decline in the late 1930s of the
reputation of Nazi Germany within the United States.

The decrease in contacts between American eugenicists and German racial hygienists was closely connected with the radicalization of
anti-Semitism in Nazi Germany. That anti-Semitism was the dominant
element of the National Socialist race ideology became clear for the
American eugenics movement when, in September 1935, the Nazi
government passed the Nuremberg laws. One of them, the Reich Citizenship Law, stipulated that only persons of "German or related

blood" could be citizens of the Reich. Jews were explicitly excluded. The other, the "Blood Protection Law," forbade marriages between Jews and "citizens of German or related blood." On November 9 and November 10, 1938, a government-sponsored pogrom against the German-Jewish population graphically illustrated the extent of discrimination against Jews in Germany. The pogrom was accompanied by a flurry of decrees that further limited Jews in nearly all aspects of their lives. They were forbidden to participate in public life, and Jewish children were no longer allowed to attend public schools. In 1939, decrees followed that limited the districts in which Jews could rent apartments, mandated forced labor, and compelled them to wear the yellow star of David.

American eugenicists became concerned that racial anti-Semitism in Germany would alter the relations between eugenicists of the two nations. Despite widespread anti-Semitism within the eugenics movement in the United States, only a few American eugenicists agreed with the degree to which Jews were discriminated against in Germany.[2] Reports in American newspapers that German racial hygienists played an important role in legitimizing Nazi anti-Semitism therefore played a strong role in changing the behavior of American eugenicists toward their German colleagues. With the increasing American criticism of the anti-Semitic policy in Nazi Germany, it became difficult even for mainline eugenicists to support Nazi race policies openly and to maintain close relationships with their German colleagues.

Nazi propagandists reacted to American criticism by arguing that ethnic minorities in the United States were treated in a similar way as were Jews in Germany. According to the Nazi view, large parts of the American public were critical of discrimination against Jews, but did not apply the same standards to all forms of ethnic racism, particularly in regard to blacks.

In 1937, the *Preussische Zeitung* claimed, under the title "The 'cruel German racial theory' and its comparisons abroad," that "liberal circles" that criticized German race laws as an "intervention in human freedom" overlooked the fact that a "state which can be seen as democratic" had its own race laws. The newspaper informed its readers that in thirty states in the United States, marriage between blacks and whites was forbidden. It also referred to the strict segregation between whites and blacks and pointed out that lynching of ethnic minorities was a phenomenon not found in Germany.[3] In a more extensive article, the same author, Wilhelm Jung, claimed that nearly all "misinterpretation, criticism, and attacks" against Nazi Germany referred to its racial measures. Compared to other nations, he viewed the United States as the only other country with extensive race legislation.[4]

In 1939, the *Nationalsozialistische Partei Korrespondenz* published an article under the headline "Double Standard in the U.S.," which claimed that America presented itself as the country of freedom, taking upon itself the task of defending humankind from the "race mania" of the authoritarian German state. The article argued that, in reality, the United States did not live up to its democratic claims. References to the lynching of blacks and the failure of an antilynching bill in the Senate were used to illustrate the "double standard."[5] Similarly, the *Berliner Börsenzeitung* reported that blacks in the United States, "in contrast to the Jews in modern Germany, knew what lynching was." The article argued,

> The Nigger would well be surprised that the white American becomes outraged at the elimination of Jews from German universities, while they do not even consider the exclusion of Negroes from many American universities.[6]

In Nazi propaganda after the mid-1930s, the United States became the main point of reference, by reason of its specific combination of ethnic and eugenic racism as well as the extent to which information on American eugenics was available in Germany.[7] In addition to the well-known books of Madison Grant and Lothrop Stoddard, Heinrich Krieger's 1936 book, *Das Rassenrecht in den Vereinigten Staaten* [*The Race Law in the United States*], provided Nazi propagandists with detailed data. The *Grossdeutscher Pressedienst* heralded Krieger's book by stating,

> [F]or us Germans it is especially important to know and to see how one of the biggest states in the world with Nordic stock already has race legislation which is quite comparable to that of the German Reich.[8]

Krieger himself defended the importance of studying the race laws in the United States by asserting that the United States was the only country besides the German Reich and South Africa that had "real race legislation."[9] Krieger placed his work under the credo:

> The central problem of today is: How are all countries of Nordic stock, especially the three leading powers of America, England, and Germany, going to learn the great theory, that they have one faith, and that up to now every injury to one of them is an injury to the others, too.[10]

Before the beginning of World War II, Nazi propagandists claimed that Germany had no interest in waging war against nations that belonged to the same "white Nordic stock." American eugenicists who believed this position reacted with surprise when the Nazis initiated aggression against nations of similar racial composition. Nazi aggression obviously strained relations between Germany and the American eugenics movements. The visits of T. U. H. Ellinger and Stoddard at the beginning of the war were already strongly affected by the difficult international situation. The complete break between German and American eugenicists occurred with the entrance of the United States into the war against Germany and Japan. After December 7, 1941, the day of the Japanese attack on Pearl Harbor, no contact between the German and American eugenics movements can be found. Many German racial hygienists were preoccupied with coordinating extensive extermination programs and the American eugenics movement suspended most of its activities for the duration of the war. The New Jersey League for Sterilization, for example, virtually ceased activity, and the American Eugenics Society limited its business to publishing *Eugenic News*.[11]

Eugenics after 1945

After World War II, members of the American Eugenics Society sought to distance themselves from their former support for Nazi race policies. The elimination of millions of Jews, Gypsies, and handicapped people had completely discredited Nazi race policies. Maurice A. Bigelow's "Brief History of the American Eugenics Society," published in *Eugenic News* in 1946, did not mention the Society's former support for Nazi attempts at race improvement. Neither did Frederick Osborn's "History of the American Eugenics Society," published in 1974.[12] Reform eugenicists' artificial distinction between favorable parts of Nazi race policy and parts that needed to be condemned or concealed influenced their self-perception after 1945. The fact that they had criticized elements of Nazism allowed them conveniently to "forget" their prior support for Nazi eugenic racism. By successfully concealing this part of their history, reform eugenicists also shaped the early historiography of eugenics, in which historians claimed that only a small and rapidly diminishing number of eugenicists on the far political Right supported Nazi policies.[13]

Other eugenicists had fewer scruples in confessing their former support for Nazi Germany. They continued to view the eugenics measures of the 1930s as exemplary, and referred with pride to the important role the United States had played in the development of this policy. As late as the 1970s, Marian S. Olden, the leading figure in the Association for Voluntary Sterilization, and Leon F. Whitney, secretary of the American Eugenics Society, proudly recalled their support for Nazi race policies.[14]

German Eugenicists and Their Relation to the United States after 1945

In the Nuremberg Doctors Trial in 1946 only a small group of German racial hygienists was accused of participating in government-sponsored massacres. In their defense, those accused referred to the acceptance of the scientific basis of their work outside Germany. This strategy was based on the claim that democratic states had provided a model for the Nazi race policy. Physicians accused of organizing the "euthanasia program" in Nazi Germany pointed to the United States to prove that elimination of "inferior elements" was not unique to Germany.[15] The 1927 United States Supreme Court decision affirming the legitimacy of eugenic compulsory sterilization in the United States was used by a German doctor as an example of the precedents for Nazi racial hygiene.[16]

Hermann Pfannmüller, a psychiatrist who, as director of the state mental hospital in Haar near Munich, was responsible for killing hundreds of mentally and physically handicapped people, explained to the court that the exterminations were

> just as legal as the regulation for prevention of transmission of hereditary disease and infection in marriage. These laws were passed during the National Socialist regime. But the ideas from which they arose are centuries old.[17]

The head of the Nazi program for the killing of the mentally handicapped, Karl Brandt, claimed before the court that the Nazi program for sterilization and elimination of "life not worthy of living" was based on ideas and experiences in the United States. In his defense, he included several works that supported his claims, such as Grant's *Passing of the Great Race,* Alexis Carrel's *Man, the Unknown,* and a study by Erich Ristow, which pointed out that Indiana's sterilization law dated back to 1907.[18]

American prosecutors at the Nuremberg Doctors Trial were not entirely unsympathetic to these arguments. Nazi racial hygienists were not tried for the forced sterilization of more than 400,000 Germans on the basis of the Law on Preventing Hereditarily Ill Progeny. Recent scholarship has shown that a group of prosecutors tried to present the mass killing of handicapped people and the experiments in concentration camps as completely separate from "genuine eugenics." They also attempted to show that American military authorities tried to recruit some of those accused of war crimes for military research.[19]

Other German racial hygienists, who were either not directly involved in the mass killings or who were able to hide their involvement, also encountered problems due to their numerous links with National Socialism. Their contacts with American colleagues helped them reestablish their position in the international scientific community shortly after the collapse of Nazi Germany.

A central figure in aiding this speedy reintegration of German scientists into the international scientific community was Hans Nachtsheim. Nachtsheim, who played an important role in genetics in Nazi Germany but did not participate in the mass killings, took over the position of director of the Kaiser Wilhelm Institute. Immediately after the end of World War II, Nachtsheim reforged close ties to his colleagues in the United States. With their help, he was able to bring geneticists with a Nazi past to the Eighth International Congress of Genetics, held in Stockholm in 1948. Nazi psychiatrists were reintegrated into the international movement at a similar speed. Leading German psychiatrists, some of whom had been closely connected to the killing of handicapped persons in Nazi Germany (such as Ernst Rüdin and Werner Villinger), were able to participate at international meetings shortly after the war.[20]

Nachtsheim tried to protect his former chief and predecessor at the Kaiser Wilhelm Institute, Freiherr von Verschuer, from accusations that he had connections to medical experiments conducted in concentration camps. Von Verschuer was accused by German physicist Robert Havemann of receiving "human material" from his assistant Josef Mengele. Before his enlistment in the S.S. in 1940, Mengele had worked under Verschuer in Frankfurt.[21] In 1942, Mengele was appointed to the Reichsarzt S.S. und Polizei in Berlin and assumed responsibility for medical experiments in concentration camps. He then contacted Verschuer, who advised him to request a transfer to Auschwitz as a "unique possibility" for racial biological research.[22] At Auschwitz, Mengele examined twins and dissected them after they were killed. He sent the results of dissections (including pairs of eyes) to Verschuer at the Kaiser Wilhelm Institute.[23] Miklos Nyiszli, doctor and prisoner at Auschwitz who worked with Mengele in preparing the specimens, confirmed this in his autobiography and claimed that Verschuer thanked Mengele for "the rare and valuable specimens."[24]

Mengele visited his professor in Berlin and was received by Verschuer's family. Shortly after the war, Verschuer destroyed all his correspondence with Mengele and denied that Mengele had ever been his assistant in Berlin or that he had ever received biological specimens from him.[25] Furthermore, Verschuer claimed that he was "openly opposed to the National Socialist race fanaticism."[26] Writing to Hermann J. Muller, Verschuer expressed his commitment to restoring the reputation of "our science." Verschuer argued that the first necessary step was to remove all those "who were not real scientists" from their positions. Referring to his own trouble, he asked Muller to support him with a letter of recommendation and lamented his "life of deprivation."[27]

Verschuer's troubles, however, did not last long. After he was classified as a "fellow traveler" in a de-Nazification process, he became professor of human genetics at Münster in 1951. Shortly thereafter he was elected president of the German Society for Anthropology.[28] Verschuer became editor of the prestigious *Zeitschrift für menschliche Vererbungs- und Konstitutionslehre* and served as a member of the editorial board of *The Mankind Quarterly*, which was later edited by Roger Pearson.[29] Verschuer's case was all too typical. Other racial hygienists who played prominent roles in Nazi Germany quickly regained influential positions. Between 1946 and 1955, for example, Fritz Lenz, Günther Just, and Heinrich Schade returned to professorships in German universities in human genetics, anthropology, or psychiatry, not in racial hygiene or eugenics.[30]

The only person who could not find a new academic position was Hans F. K. Günther. Although the American Society for Human Genetics helped with the nomination of Günther as one of its foreign members, at least to reestablish his academic self-esteem, Günther never could get a new position as a professor in Germany.[31] His thinking was too abstruse, and his theories were too discredited. This, however, did not prevent Günther from publishing more academic books, even if he sometimes used his pseudonyms: "Ludwig Winter" or "Heinrich Ackermann."[32] In the late 1950s, Günther found a new group of intellectuals with whom he could celebrate his "Nordic" thoughts: The Northern League, established in 1958 by Roger Pearson.[33]

That the ideology of many former German eugenicists and racial hygienists did not change is revealed by the attempts of a group of eugenicists who were not involved in the mass killing to reestablish a eugenic sterilization program. They used references to sterilizations in the United States and other countries to separate eugenic measures from National Socialism. As early as 1949, Karl Bonhoeffer, a famous psychiatrist at the Charité Hospital in Berlin, attempted to revive the sterilization program. Citing the absence of discrimination against ethnic minorities in the German law, the positive reaction toward the Nazi law abroad, and the support of the 1935 International Congress for Criminal Law in Berlin, Bonhoeffer pleaded for the reestablishment of eugenics policy in Germany.[34]

Hans Harmsen, the chief ideologist of the eugenicists in the Protestant church, and Hans Nachtsheim voiced similar thoughts. They argued that sterilization laws had value independent of their implementation under National Socialism. In their view, therefore, it would be appropriate to continue the practice of sterilization in a nontotalitarian society.[35]

Conclusion

> I think it would be a great mistake to identify eugenic sterilization solely with the Nazi ideology and to dismiss the problem simply because we dislike the present German regime and its methods. . . . The problem is serious and acute, and we shall be forced to pay attention to it sooner or later.[1]
>
> Henry Sigerist, science historian at Johns Hopkins University in 1943

Attempts to separate eugenics from the Nazi program of race improvement were only partially successful. The personal and ideological links between eugenics and mass sterilization and extermination were too obvious to be overlooked. Socialist eugenicists, who opposed Nazi race policies, distinguished themselves by avoiding the word *eugenics*. Instead, eugenicists such as Hermann Muller introduced terms like *genetic load* and *cost of selection*.[2] Reform eugenicists who had supported parts of the Nazi race policies also backed away from the term *eugenics* after it had become tainted due to Nazi abuses. By 1954, the British *Annals of Eugenics* was renamed the *Annals of Human Genetics*; in 1969, *Eugenics Quarterly,* the successor of *Eugenic News,* was renamed the *Journal of Social Biology*. The pride with which scientists in the 1910s, 1920s, and 1930s referred to themselves as eugenicists had evaporated. After World War II, eugenicists described themselves as "population scientists," "human geneticists," "psychiatrists," "sociologists," "anthropologists," and "family politicians" in an attempt to avoid eugenics terminology.

Eugenicists tried to separate themselves from the legacy of the Holocaust and the ideology of Nordic superiority by eliminating ethnic racism from the official agenda of eugenics societies. This move helped them regain acceptance in the scientific community in the 1940s and 1950s, when the categorization of races as "inferior" and "superior" had become widely discredited among scientists. Birthright, Inc., the successor of the New Jersey League for Sterilization, declared that its purpose was "to foster all reliable and scientific means for improving the human race."[3] Although some of its members had been strong promoters of ethnic racism, Birthright, Inc., did not advocate sterilization on the basis of ethnicity.

The Pioneer Fund, however, tried to keep ethnic racism alive within eugenics organizations. Wickliffe Draper's foundation, which continued to support the American Eugenics Society and Birthright, Inc., after 1945, remained a bastion of mainline eugenics. During a lunch with Frederick Osborn on October 26, 1954, Draper offered to guarantee his full support of the Eugenics Society for a minimum of five years if the Society would assume a position more along the lines of his thinking. Such an approach had to include "measures for establishing racial homogeneity in the United States." Osborn rejected the offer; Draper's belief, so he argued, had "at present" no basis in scientific findings.[4]

Pioneer Fund grantees Roger Pearson, Hans J. Eysenck, Arthur Jensen, Robert A. Gordon, J. Philippe Rushton, and Linda Gottfredson are today engaged in providing scientific findings for genetic differences between races. Thus, a Fund that was founded by supporters of Hitler's policies against ethnic minorities and handicapped people and that provided money for introducing Nazi race propaganda into the United States still sponsors research that has striking similarities to earlier studies that provided the scientific basis for Nazi measures.

The bureaucratized mass killing of millions of human beings because of "racial inferiority" is obviously historically unique, but continues to serve as a sobering reminder of what can happen when eugenic and ethnic racism are combined in a comprehensive program of race improvement. Unlike their German colleagues, American scientists did not participate in the selection of tens of thousands of handicapped people for the Nazi gas chambers. Nevertheless, the involvement of American eugenicists with Nazi policies reveals that the ideology of race improvement that was at the root of the massacres was by no means limited to German scientists.

Notes

Introduction

1. An interim report about my research at Bethel was published at the University Bielefeld: Stefan Kühl, *Bethel zwischen Anpassung und Widerstand: Die Auseinandersetzung der von Bodelschwinghschen Anstalten mit der Zwangssterilisation und den Kranken- und Behindertenmorden im Nationalsozialismus* (Bielefeld: AStA der Universität Bielefeld, 1990). Copies are available at the Hauptarchiv der von Bodelschwinghschen Anstalten, Bethel bei Bielefeld. Over the course of several years, the von Bodelschwinghschen Anstalten hindered my access to important sources that could have clarified some of my questions regarding the institution. For general information about sterilization and killing of the mentally handicapped in Protestant institutions, see: Kurt Nowak, *"Euthanasie" und Sterilisierung im "Dritten Reich": Die Konfrontation der evangelischen und katholischen Kirche mit dem "Gesetz zur Verhütung erbkranken Nachwuchses" und der "Euthanasie"-Aktion* (Halle: Niemeyer, 1977); Ernst Klee, *"Euthanasie" im NS-Staat: Die "Vernichtung lebensunwerten Lebens"* (Frankfurt a.M.: Fischer, 1983); Jochen-Christoph Kaiser, "Innere Mission und Rassenhygiene: Zur Diskussion im Centralausschuss für Innere Mission 1930–1938," *Lippische Mitteilungen,* 55 (1986): 197–217 and *Sozialer Protestantismus im 20. Jahrhundert: Beiträge zur Geschichte der Inneren Mission 1918–1934* (Munich: Oldenbourg, 1989).
2. Barry Mehler pointed out that historians Mark H. Haller and Kenneth M. Ludmerer based their assessment in part on interviews conducted with Frederick Osborn after 1945. Mark H. Haller, *Eugenics: Hereditarian Attitudes in American Thought* (New Brunswick: Rutgers University Press, 1963): 174, note 27; and Kenneth M. Ludmerer, *Genetics and American Society* (Baltimore, London: Johns Hopkins University Press, 1972): 174, note 39 on page 239. See: Barry Mehler, "A History of the American Eugenics Society, 1921–1940," diss., University of Illinois, 1988, 224.
3. Maurice A. Bigelow, "Brief History of the American Eugenics Society," *Eugenic News,* 31 (1946): 49–51; Frederick Osborn, "History of the American Eugenics Society," *Social Biology,* 21 (1974): 115–26.
4. Haller, *Eugenics,* 182.
5. Ludmerer, *Genetics,* 117.
6. Carl L. Bajema, ed., *Eugenics: Then and Now* (Stroudsburg: Hutchinson & Ross, 1976): 14, note 20. See: Mehler, *History,* 226.

7. Garland E. Allen and Barry Mehler in "Sources in the Study of Eugenics #1: Inventory of the American Eugenics Society Papers," *Mendel Newsletter, Archival Resources for the History of Genetics and Allied Sciences*, 14 (1977): 9–15. Laughlin's connections to Nazism had been discussed previously in: Frances J. Hassencahl, "Harry H. Laughlin: 'Expert Eugenics Agent' for the House Committee on Immigration and Naturalization, 1921 to 1931," diss., Case Western Reserve University, 1970. When I speak of "Nazi racial hygienists," I mean the German racial hygienists who supported Nazi race policies, participated in their implementation, or provided Nazi race ideology with scientific legitimation.

8. Allen Chase, *The Legacy of Malthus: The Social Costs of the New Scientific Racism* (New York: Alfred A. Knopf, 1977): 15.

9. John David Smith, *Minds Made Feeble: The Myth and Legacy of the Kallikaks* (Rockeville: Aspen Systems Corporation: 1985); Thomas M. Shapiro, *Population Control Politics: Women, Sterilization and Reproductive Choice* (Philadelphia: Temple University Press, 1985).

10. Lanny Lapon, *Mass Murder in White Coats: Psychiatric Genocide in Nazi Germany and the United States* (Springfield: Psychiatric Genocide Research Institute, 1986).

11. Daniel J. Kevles, *In the Name of Eugenics: Genetics and the Uses of Human Heredity* (Berkeley, Los Angeles: University of California Press, 1986): 347. Kevles' goal of understanding eugenics in the United States and Great Britain in relation to the developments in Germany is stated on page x. See also: Barry Mehler, review of *In the Name of Eugenics*, *Journal of Social History*, 20 (1987): 616–19.

12. Mehler, *History*, 223–68. See also: Barry Mehler, "Eliminating the Inferior," *Science for the People*, 21.6 (1987): 14–18.

13. Stephen Trombley, *The Right to Reproduce: A History of Coercive Sterilization* (London: Weidenfeld & Nicolson, 1988).

14. The California eugenics movement—the most active of any state—has been neglected in the historiography. One exception is an article by Melanie Fong and Larry O. Johnson, "The Eugenics Movement: Some Insight into the Institutionalization of Racism," *Issues in Criminology*, 9 (1974): 89–115.

15. Robert Proctor, *Racial Hygiene: Medicine under the Nazis* (Cambridge and London: Harvard University Press, 1988): 96–101; "Eugenics among Social Sciences: Hereditarian Thought in Germany and the United States," *The Estate of Social Knowledge*, ed. JoAnne Brown and David K. van Keuren (Baltimore: Johns Hopkins University Press, 1991): 175–208.

16. Gisela Bock, *Zwangssterilisation im Nationalsozialismus: Studien zur Rassenpolitik und Frauenpolitik* (Opladen: WDV, 1986); Peter Weingart, Jürgen Kroll, and Kurt Bayertz, *Rasse, Blut und Gene: Geschichte der Eugenik und Rassenhygiene in Deutschland* (Frankfurt a.M.: Suhrkamp, 1988).

17. Studies focusing on the eugenics movement as an ideological predecessor to the race policies of Nazi Germany include: George L. Mosse, *The Crisis of German Ideology: Intellectual Origins of the Third Reich* (New York: Grosset and Dunlap, 1964); George L. Mosse, *Toward the Final Solution: A History of European Racism* (London: J. M. Dent, 1978); Günther Altner, *Weltanschauliche Hintergründe der Rassenlehre des Dritten Reiches: Zum Problem einer umfassenden Anthropologie* (Zurich: EVZ, 1968); Daniel Gasman, *The Scientific Origins of*

National Socialism: Social Darwinism in Ernst Haeckel and the German Monist Leaque (New York: American Elsevier, 1971); Patrick von Mühlen, *Rassenideologien: Geschichte und Hintergründe* (Berlin, Bad Godesberg: J. H. W. Dietz, 1977). Studies that attempt to separate the history of social Darwinism and the early eugenics movement from the later radicalization of Nazi racial hygiene include: Sheila F. Weiss, *Race Hygiene and National Efficiency: The Eugenics of Wilhelm Schallmayer* (Berkeley, Los Angles, London: University of California Press, 1987); Alfred Kelly, *The Descent of Darwin: The Popularization of Darwinism in Germany, 1860–1914* (Chapel Hill: The University of North Carolina Press, 1981); Loren R. Graham, "Science and Values: The Eugenics Movement in Germany and Russia in the 1920s," *American Historical Review*, 82 (1977): 1135–64.

18. See for example: Bock, *Zwangssterilisation*; Weingart, Kroll and Bayertz, *Rasse*. For the United States, see: Mehler, *History*. For Canada, see: Angus McLaren, *Our Own Master Race: Eugenics in Canada, 1885–1945* (Toronto, Ontario: McClelland & Steward, 1990). For the eugenics movements in Latin America, see: Nancy L. Stepan, *The Hour of Eugenics: Race, Gender and Nation in Latin America* (Ithaca: Cornell University Press, 1991). For France, see: William H. Schneider, *Quality and Quantity: The Quest for Biological Regeneration in Twentieth-Century France* (Cambridge: Cambridge University Press, 1990). I use the word *transnational* to define cooperation across national borders. *International* describes the stabilization of transnational contacts within institutions, as well as shared ideology.

19. In this book I use *racial hygiene* and *eugenics* as synonyms. On an international level, eugenicists and racial hygienists used these terms interchangeably. For the first studies to attempt a comparative history of eugenics, see: *The Wellborn Science: Eugenics in Germany, France, Brazil and Russia*, ed. Mark B. Adams (New York, Oxford: Oxford University Press, 1990); Paul Weindling, "The 'Sonderweg' of German Eugenics: Nationalism and Scientific Internationalism," *The British Journal of the History of Science*, 22 (1989): 321–33; Graham, *Science and Value*; Jonathan Harwood, "National Styles in Science: Genetics in Germany and the United States between the Wars," *ISIS*, 78 (1987): 390–414.

20. On the eugenics movement in socialist states: Graham, *Science*; Mark B. Adams, "Eugenics in Russia 1900–1940," *The Wellborn Science*, ed. Adams, 153–216. On the eugenics movement under fascist regimes: R. Alvarez-Peláez, "El Instituto de Medicina Social: Primeros Intentos de Institucionalizar la Eugenesia (en Espana)," *Acta Ill Congreso Sociedad Espanola de Historia de las Ciencias*, Octubre 1984; "Introduction al Estudio de la Eugenesia Espanola (1900–1936)," *Quipu* 2 (1985): 95–122; Claudio Pogliano, "Scienza e Stirpe: Eugenica in Italia (1912–1939)," *Passato e Presente*, 5 (1984): 61–79. The first comparative work of eugenics in fascist countries is: Paul Weindling, "Fascism and Population Policies in Comparative European Perspective," *Population, Resources and the Environment: The Interplay of Science, Ideology and Intellectual Traditions*, eds. M. Teitelbaum and J. Winter (Cambridge: Cambridge University Press, 1988): 102–20.

21. Bilateral analysis can contribute to the discussion as to whether German racial hygiene fulfilled a *Sonderweg* in the history of national eugenics movements. Similarities between German and non-German eugenicists would suggest that a

Sonderweg could only have taken place in the state bureaucracy and in the medical profession, not in the movement that was responsible for the ideological basis. A full-scale discussion of a German *Sonderweg* regarding the implementation of racism as the basic principle of state policy after 1933 has not yet taken place. See: Paul Weindling, *Health, Race and German Politics between National Unification and Nazism, 1870–1945* (Cambridge: Cambridge University Press, 1989); Weingart, Kroll, Bayertz, *Rasse*; Hans Walter Schmuhl, *Rassenhygiene, National-sozialismus, Euthanasie: Von der Verhütung zur Vernichtung 'lebensunwerten Lebens' 1890–1945* (Göttingen: Vandenhoeck & Ruprecht, 1987).

Chapter 1

1. This statement was prepared by a committee of experts on race and racial prejudice that met at UNESCO House, Paris, from September 18 to 26, 1967, and was reprinted in *Race, Science and Society*, ed. Leo Kuper (Paris: The UNESCO Press, 1975): 360–64.
2. I use "scientific racism" in a sociological sense to describe tendencies in science. Underlying my understanding of scientific racism is the concept of "science" as socially constructed knowledge. Consequently, I use the word *pseudoscientific* only to describe science that is *consciously* manipulated to support desired scientific, social, or political conclusions. Since the 1940s eugenicists have restrained themselves more and more from using the words "superior" and "inferior." I claim, however, that the sense of the word "inferiority" is kept and only paraphrased when scientists refer to ethnic groups as less intelligent due to genetic reasons.
3. Roger Pearson, *Race, Intelligence and Bias in Academe*, intro. Hans J. Eysenck (Washington: Scott-Townsend, 1991): 13.
4. Michael Billig, *Psychology, Racism and Fascism* (Birmingham: Searchlight, 1979): 7.
5. Paul Valentine, "The Fascist Spectre Behind the World Anti-Red League: Neo-Fascist Forces Astir," *Washington Post* (May 28, 1978). See: Barry Mehler, "The New Eugenics: Academic Racism in the U.S. Today," *Science for the People*, 15.3 (1983): 19.
6. Pearson wrote several anthropological textbooks, including *An Introduction to Anthropology* (New York and London: Holt, Rinehart and Winston, 1974); *Anthropological Glossary* (Malabar: Krieger Publishing, 1985). In the 1960s, he taught at Queens College in Charlotte, North Carolina, and at the University of Southern Mississippi in Hattiesburg before becoming dean of academic affairs at Montana College in Butte. See: Billig, *Psychology*, 18.
7. Reagan specifically noted Pearson's service "in bringing to a wide audience the work of leading scholars who are supportive of a free enterprise economy, a firm and consistent foreign policy and a strong national defense." The letter was uncovered by researcher Russell Bellant and reprinted in Barry Mehler, "Eugenics: Racist Ideology Makers," *Guardian* (August 22, 1984).
8. Pearson, *Race*, 56. In addition to Rushton and Gordon, this group includes scholars like Arthur Jensen and William Shockley.
9. Pearson, *Race*, 56.
10. Francis Galton, *Inquiries into Human Faculty* (London: Macmillan, 1883): 24. American eugenicist Raymond Pearl defined eugenics as

> [T]he science which deals with all influences that improve the inborn quali-
> ties of a race, also with those that develop them to the utmost advantage; and
> it embodies the study of agencies under social control that may improve or
> impair the racial qualities of future generations.

The opposite of *eugenics* is *dysgenics,* and is defined by Pearl as the study of social processes that lead to a decline in the genetic quality of humankind. Raymond Pearl, "Breeding Better Men," *World's Work,* 15 (1908): 9819. From a historical viewpoint, Sheila F. Weiss defines eugenics as "a political strategy denoting some sort of social control over reproduction. In the interest of 'improving' the hereditary substrata of a given population, this supposed science seeks to regulate human procreation by encouraging the fecundity of the allegedly genetically superior groups ('positive eugenics') and even prohibiting so-called inferior types from having children ('negative eugenics')." Weiss, *Race Hygiene,* 1.

11. Galton, *Inquiries,* 24.
12. Pearson, *Race,* 57. I detail Pearson's description of eugenics in order to clarify the tradition in which he locates himself and his collegues. His description is accurate for the large majority of eugenicists, but he neglects to mention a group that I describe as socialist eugenicists. See Chapter 7 for more information.
13. Pearson, *Race,* 57–58.
14. For example, Pearson mentions a case in 1971 in which delegates of the American Psychological Association in Washington, D.C., accused Shockley of "racism and of promoting 'fascist' ideas associated with Nazi Germany." Pearson refers to a report in *The Sacramento Union* (November 23, 1971). See: *Race,* 191.
15. Hans J. Eysenck's works include *Race, Intelligence and Education* (London: Temple Smith, 1971) and *The Inequality of Man* (London: Temple Smith, 1973). His introduction to the German translation of *The Inequality of Man* is also interesting. Hans J. Eysenck, *Die Ungleichheit der Menschen* (Munich: Goldmann, 1978): 9–38. For a laudatory biography see: H. B. Gibson, *Hans Eysenck: The Man and His Work* (London: Peter Owen, 1981). A critical appraisal is: Michael Billig, "Pofessor Eysenck's political psychology," *Patterns of Prejudice,* 13.5 (1979): 9–16.
16. Hans J. Eysenck, "Science and Racism," introduction to Pearson, *Race,* 40.
17. Information about the figures of financial support by the Pioneer Fund is drawn from federal tax returns, available at the Foundation Center, Washington: Pioneer Fund.
18. Certificate of the Pioneer Fund, February 27, 1937, signed by Harry H. Laughlin, Frederick Osborn, Wickliffe Draper, Malcolm Donald, and Vincent R. Smalley. Laughlin Papers, Missouri State University, Kirksville.
19. I thank Barry Mehler for providing me with information about the activities of the Pioneer Fund. The information presented is from his articles: *New Eugenics;* "Foundation for Fascism: the New Eugenics Movement in the United States," *Patterns of Prejudice,* 23.4 (1989): 17–25; "Study Can't Hide Taint of Eugenics," *The Jewish News* (September 1, 1989).
20. Grace Lichtenstein, *The New York Times* (December 11, 1967). In Mehler, *Foundation,* 21–22. For information about Scott's career as chairman of the Iowa State Advisory Committee to the U.S. Civil Rights Commission see: Barry Mehler, "Ralph Scott's curious career: rightist on the rights panel," *Nation* (May 7, 1988):

40–41 and Susan Birnbaum, "Iowa Civil Rights Adviser Resigns Following Book Review Controversy," *Jewish Telegraphic Agency* (July 14, 1988).

21. Pearson, *Race,* 141. Pearson's forty-page tribute to Jensen is based on material that Jensen provided for Pearson. See: Pearson, *Race,* 15.

22. Arthur Jensen, "How Much Can We Boost I.Q. and Scholastic Achievement?" *Harvard Educational Review,* 39 (1969): 1–123. His theory is elaborated in his books *Genetics and Education* (New York: Harper & Row, 1972) and *Educability and Group Differences* (New York: Harper & Row, 1973). For a summary of his main thesis, see: *Newsweek* (March 31, 1969): 84. An interview with Jensen is published under the title "Rasse und Begabung" in *Nation Europa,* 25.9 (1975): 19–28.

23. From a letter mailed by Shockley to members of the National Academy of Sciences, April 16, 1970, on the letterhead of the Foundation for Research and Education on Eugenics and Dysgenics, with a cover letter by Shockley. Quoted in Chase, *Legacy,* 482. Pearson's praise of Shockley's plan is in *Race* on page 192.

24. Quoted in Chase, *Legacy,* 484.

25. Mehler, *The New Eugenics,* 19.

26. J. Philippe Rushton, "Evolutionary Biology and Heritable Traits," Paper presented at the Symposium on Evolutionary Theory, Economics, and Political Science, American Association for the Advancement of Science (San Francisco, January 19, 1989). J. Philippe Rushton, "Sir Francis Galton, epigenetic rules, genetic similarity theory, and human life-history analysis," *Journal of Personality,* 58 (1990): 117–40. Rushton has also published in Pearson's *The Mankind Quarterly*: J. Philippe Rushton, "Evolution, Altruism and Genetic Similarity Theory," *Mankind Quarterly,* 27 (1987): 379–95.

27. J. Philippe Rushton, "Race Differences in Behavior: A Review and Evolutionary Analysis," *Personality and Individual Differences,* 9 (1988): 1009–24; "The Reality of Racial Differences: A Rejoinder with New Evidence," *Personality of Individual Differences,* 9 (1989): 1035–40. See references to both these articles in: Pearson, *Race,* 216.

28. J. Philippe Rushton and Anthony F. Bogaert, "Population Differences in Susceptibility to AIDS: An Evolutionary Analysis," *Social Science and Medicine,* 28 (1989): 1211–20. See also: Rushton, "Race Differences in Sexuality and Their Correlates: Another Look and Physiological Models," *Journal of Research in Personality,* 23 (1989): 35–54, and the heated discussion between Rushton, Charles Leslie, C. Owen Lovejoy, Glenn D. Wilson, and Peter J. M. McEwan, "Scientific Racism: Reflections on Peer Review, Science and Ideology," *Social Science and Medicine,* 31 (1990): 891–905.

29. Pearson, *Race,* 236. See also p. 15 for references to the cooperation of Rushton with Pearson.

30. For example, in October 1971 Gordon submitted his manuscript, "An Explicit Estimation of the Prevalence of Commitment to a Training School, to Age 18, by Race and by Sex," to the *Journal of the American Statistical Association* after it and a similar paper had been rejected by major sociological and criminological journals. Two years later, a revised version of his paper was printed by the *Journal of the American Statistical Association,* 68 (1973): 547–53.

31. Robert A. Gordon, "Examining Labeling Theory: The Case of Mental Retardation," *In the Labeling of Deviance: Evaluating a Perspective,* ed. Walter R. Gove

(Beverly Hills: Sage, 1975): 83–146; "Comment on 'Delinquency, Sex, and Family Variables' by Andrew," *Social Biology*, 24 (1977): 337; "Research on IQ, Race, and Delinquency: Taboo or Not Taboo?" *Taboos in Criminology*, ed. Edward Sagarin (Beverly Hills: Sage, 1980): 37–66; "The Black–White Factor is g," *Behavioral and Brain Sciences*, 8 (1985): 229–31; "Jensen's Contributions Concerning Test Bias: A Contextual View," *Arthur Jensen: Consensus and Controversy*, eds. Sohan Modgil and Celia Modgil (New York: Falmer Press, 1987): 77–154; "SES versus IQ in the Race–IQ–Deliquency Model," *International Journal of Sociology and Social Policy*, 7.3 (1987): 30–96.

32. Robert A. Gordon, "Crime and Cognition: An Evolutionary Perspective," *Proceedings of the II International Symposium on Criminology*, vol. 4 (Sao Paulo: International Center for Biological and Medico-Forensic Criminology, 1975): 7–55.

33. Robert A. Gordon, "IQ-Commensurability of Black–White Differences in Crime and Delinquency," Paper presented at the annual meeting of the American Psychological Association, Washington, D.C., August 24, 1986. In this paper Gordon reviewed a selection of studies concerning differences in IQ rates of blacks and whites carried out between 1921 and 1980.

34. Gordon in a letter to the University of Delaware, March 30, 1990, quoted in Pearson, *Race*, 13. See: Linda S. Gottfredson, Jan H. Blits, "Employment Testing and Job Performance," 98, Winter (1990): 19–25. See also: Nathan Brody, Michael C. Corballis, Linda S. Gottfredson, William Shockley, Arthur R. Jensen, "Commentary on Arthur R. Jensen (1955), The Nature of Black–White Differences on Various Psychometric Tests: Spearman's Hypothesis," *Behavioral and Brain Science*, 10 (1987): 507–18.

35. According to Pearson, *Race*, 279.

36. President of the Pioneer Fund, Harry Weyher, remarked that the bulk of Pioneer Fund money goes to support studies of twins raised separately. Jack Anderson and Dale van Atta, "Pioneer Fund's Controversial Projects," *The Washington Post*, November 16, 1989.

37. Thomas J. Bouchard, Auke Tellegen, David T. Lykken, Kimerly J. Wilcox, "Personality Similarity in Twins Reared Apart and Together," *Journal of Personality and Social Psychology*, 54 (1988): 1031–39; Richard D. Arvey, Thomas J. Bouchard, Nancy L. Segal, and Lauren M. Abraham, "Job Satisfaction: Environmental and Genetic Components," *Journal of Applied Psychology*, 74 (1989): 187–92; Thomas J. Bouchard and Matthew McGue, "Genetic and Rearing Environmental Influences on Adult Personality: An Analysis of Adopted Twins Reared Apart," *Journal of Personality*, 58 (1990): 263–93; Thomas J. Bouchard, David T. Lykken, Matthew McGue, Nancy L. Segal, and Auke Tellegen, "Sources of Human Psychological Differences: The Minnesota Study of Twins Reared Apart," *Science*, 250.4978 (1990): 223–28. See also: "Zweites Ich," *Der Spiegel*, 44.16 (1990): 248–50.

38. Pearson, *Race*, quotes on pages 297–300.

39. Other institutions that have received or currently receive money from the Pioneer Fund include the Federation for American Immigration Reform, Alexandria, Virginia; Institute for the Study of Educational Differences, Orlinda, California; the Atlas Economic Research Fund, Fairfax, Virginia; National Hemophilia Foundation, New York; Pearson's Council for Social and Economic Studies; Charles

Darwin Research Institute in London, Canada; University of Pennsylvania in Philadelphia; Regents University of California, Santa Barbara, California; Tel Aviv University in Israel; the University of Illinois in Champaign; and the Hoover Institution of Stanford University in California. Pearson used the names of prominent Pioneer Fund recipients to justify the fact that nearly all of the scientists whom he tried to defend against the reproaches of racism were or are protégés of the Pioneer Fund. See: Pearson, *Race,* 256.

40. Robert A. Gordon, "Universities Violated Academic Principles in Pioneer Fund Ban," *The Review* (University of Delaware) (October 29, 1991).

41. Gordon quotes Huey Long's reply to the question of whether fascism would come to America, to which Long replied "Yes, but it will be called antifascism."

42. Eysenck, *Science,* 51.

43. Gibson, *Hans Eysenck,* 234.

44. Harry Weyher in a letter to *American Jewish World,* quoted from: *The Jewish Week,* July 14, 1989.

Chapter 2

1. Feilchenfeld, "Die Bestrebung der Eugenil in den Vereinigten Staaten von Nordamerika und ihre Übertragung auf deutsche Verhältnisse," *Medizin-Reform,* 21 (1913): 477–82 (my translation). Feilchenfeld's first name is not known; he appears to have played a minor role in the eugenics movement.

2. "Der erste internationale Kongress für Rassenverbesserung," *Berliner Tageblatt,* July 25, 1912.

3. *Fortpflanzung, Vererbung, Rassenhygiene: Katalog der Gruppe Rassenhygiene der Internationalen Hygiene Ausstellung 1911 in Dresden,* eds. Max von Gruber and Ernst Rüdin (Munich: Lehmann, 1911). See also: Jürgen Kroll, "Zur Entstehung einer naturwissenschaftlichen und sozialpolitischen Bewegung: Die Entwicklung der Eugenik/Rassenhygiene bis zum Jahre 1933," diss., University of Tübingen, 1983, 113; Weindling, *Health,* 144.

4. Eugenics Education Society, *The First International Eugenic Congress, London* (London: Eugenics Education Society, n.d.): 4. See also: *Problems in Eugenics: Papers Communicated to the First International Eugenics Congress held at the University of London, July 24th to 30th, 1912* (London: Eugenics Education Society, 1912).

5. Berliner Gesellschaft für Rassenhygiene, *Die Leistung der Amerikaner auf rassenhygienischem Gebiete* (Berlin: Berliner Gesellschaft für Rassenhygiene, 1917 or 1918), in: Bundesarchiv Koblenz (abbr. BAK), R 86/2371, Bd. 1.

6. Some of Hoffmann's articles for the general public appeared in the *Berliner Tageblatt* in 1913 and 1914. See: Bundesarchiv Potsdam (abbr. BAP), 49–01 REM 3198. On February 17, 1914, the *Berliner Tageblatt* ran an article entitled "Rassenhygiene in Amerika" concerning a talk Hoffmann presented before the Berlin Society for Racial Hygiene.

7. Géza von Hoffmann, *Rassenhygiene in den Vereinigten Staaten von Nordamerika* (Munich: Lehmann, 1913): 14.

8. Hoffmann, *Rassenhygiene,* 15.

9. Hoffmann, *Rassenhygiene,* 15. Ross first used the term *race suicide* in an address before the American Academy of Political and Social Science. Ross, "The Causes

of Race Superiority,'' *Annals of the American Academy of Political and Social Science*, 18 (1901): 85–88. See: Barry Mehler, ''John R. Commons,'' M.A. thesis, College of the City University of New York, 1972, 150; John Higham, *Strangers in the Land: Patterns of American Nativism, 1860–1924* (New Brunswick: Rutgers University Press, 1963): 146.

10. Hoffmann, *Rassenhygiene*, 67–68. Marriage and sexual intercourse between blacks and white was forbidden in thirty-two states.

11. Hoffmann, *Rassenhygiene*, 69.

12. Research by historian G.J. Baker-Benfield has shown that ''female castration'' was first performed already in 1872 and flourished between 1880 and 1900. Baker-Benfield, *The Horrors of the Half-Known Life: Male Attitudes Toward Women and Sexuality in Nineteenth-Century America* (New York: Harper Colophon, 1976), 121.

13. Charles Woodruff, ''Climate and Eugenics,'' *American Breeders Association, Proceedings of Annual Meetings*, 6 (1910): 122, cited in: Hoffmann, *Rassenhygiene*, 114.

14. For a commemoration of Fritz Lenz, see: ''Zum 100. Geburtstag von Fritz Lenz,'' *Neue Anthropologie*, 15 (1987): 21.

15. I use the term *Anglo-American* instead of the term *Anglo-Saxon*, which has a racial connotation. I thank Gisela Bock for clarifying my use of language in this case.

16. Fritz Lenz in *Archiv für Rassen- und Gesellschaftsbiologie*, 10 (1913): 249–52. Another review is ''Rassenhygiene in den Vereinigten Staaten,'' *Berliner Tageblatt*, May 2, 1914.

17. Géza von Hoffmann, ''Die rassenhygienischen Gesetze des Jahres 1913 in den Vereinigten Staaten,'' *Archiv für Rassen- und Gesellschaftsbiologie*, 11 (1914): 21–32.

18. Géza von Hoffmann, ''Das Sterilisierungsprogramm in den Vereinigten Staaten von Nordamerika,'' *Archiv für Rassen- und Gesellschaftsbiologie*, 11 (1914): 184. Hoffmann referred to Harry H. Laughlin, *The Legal, Legislative and Administrative Aspects of Sterilization: Report of the Committee to Study and to Report on the Best Practical Means of Cutting off the Defective Germ-plasm in the American Population* (Cold Spring Harbor: Eugenics Record Office, 1914). See also: Hoffmann, ''Rassenhygienische Jahresversammlung in den Vereinigten Staaten von Nordamerika,'' *Archiv für Rassen- und Gesellschaftsbiologie*, 10 (1913): 829–30.

19. *Eugenics Review*, 11 (1920): 249; *Eugenic News*, 4 (1919): 88.

20. Bluhm was briefly engaged to Alfred Ploetz. Ploetz sought a 'suitable healthy, Germanic wife who shared his scientific ideals. He fell in love with her during an anatomical dissection, but he later married Pauline Rüdin, a sister of one of the leading eugenicists of the time, Ernst Rüdin. Ploetz and Bluhm nevertheless remained lifelong friends. See: Weindling, *Health*, 74.

21. Davenport to Agnes Bluhm, August 8, 1921, Davenport Papers: Bluhm, American Philosophical Society (abbr. APS), Philadelphia.

22. Baur to Davenport, May 1, 1923, Davenport Papers: Baur.

23. Lenz in a letter to Mjöen. Quoted in a letter from Davenport to Lenz, from July 13, 1923, Davenport Papers: Lenz.

24. Lenz to Davenport, August 8, 1923, Davenport Papers: Lenz.

25. *Archiv für Rassen- und Gesellschaftsbiologie*, 16 (1924): 458.

26. Fritz Lenz, "Eugenics in Germany," trans. Paul Popenoe, *Journal of Heredity*, 15 (1924): 223. The original German admiration for Anglo-American eugenics in general evolved into a special admiration for the United States in particular, primarily because of the fact that British eugenicists stressed the voluntary and educational aspects of eugenics to a much stronger degree.

27. Reinhold Müller, "Geschichtliches über die Beziehungen von Vererbungslehre und Rassenkunde," *Ziel und Weg*, 2 (1934): 24, quoted in Proctor, *Racial Hygiene*, 98.

28. Paul Popenoe, "Rassenhygiene (Eugenik) in den Vereinigten Staaten," trans. Fritz Lenz, *Archiv für Rassen- und Gesellschaftsbiologie*, 15 (1923/1924): 184–86. See also a note by Popenoe about the "remarkably little divergence" within the international eugenics movement in the *Journal of Heredity*, 13 (1922): 21.

29. Roswell H. Johnson, "International Eugenics," diss. University of Pittsburgh, 1934, 186. Paul Popenoe and Roswell H. Johnson, *Applied Eugenics* (New York: Macmillan 1920, second edition 1933, third edition 1939).

30. The central importance of the Kaiser Wilhelm Institute for Psychiatry for Nazi race policies is stressed in Hans Hartmann, "Die deutsche erbbiologische Forschung: Zum 25jährigen Bestehen der Kaiser-Wilhelm-Gesellschaft zur Förderung der Wissenschaft," *Der Erbarzt*, 3 (1936): 3–8.

   For general information about the Institute, see: Matthias M. Weber, "Die Entwicklung der Deutschen Forschungsanstalt für Psychiatrie in München zwischen 1917 und 1945," *Sudhoffs Archiv*, 75 (1991): 74–89. Concerning the KWI for Anthropology, Human Heredity, and Eugenics, see: Paul Weindling, "Weimar Eugenics: The Kaiser Wilhelm Institute for Anthropology, Human Heredity, and Eugenics in Social Context," *Annals of Science*, 42 (1985): 303–18; Peter Weingart, "German Eugenics between Science and Politics," *Osiris*, ns, 5 (1989): 260–82; Anna Bergmann, Gabriele Czarnowski, and Annegret Ehmann, "Menschen als Objekte humangenetischer Forschung und Politik im 20. Jahrhundert: Zur Geschichte des Kaiser-Wilhelm-Instituts für Anthropologie, menschliche Erblehre und Eugenik in Berlin-Dahlem (1927–1945)," *Der Wert des Menschen: Medizin in Deutschland 1918–1945*, ed. Ärztekammer Berlin (Berlin: Edition Hentrich, 1989): 121–42.

   Concerning the role of the Rockefeller Foundation in supporting German racial hygiene, see: Paul Weindling, "From Philanthropy to International Science Policy: The Rockefeller Funding of Biomedical Sciences in Germany 1920–1940," *Science, Politics and the Public Good: Essays in Honor of Margret Gowing*, ed. Nicolaas Rupke (Basingstoke: Macmillan, 1988): 119–40; Bernhard Schreiber, *Die Männer hinter Hitler: Eine deutsche Warnung an die Welt* (Stuttgart: n.p., 1972): 109–10

31. Klaus-Dieter Thomann, "Otmar Freiherr von Verschuer-ein Hauptvertreter der faschistischen Rassenhygiene," *Medizin im Faschismus*, eds. Achim Thom and Horst Spaar (Berlin: VEB Verlag Volk und Gesundheit, 1985): 38–41; Weindling, *Health*, 314.

32. Letter from February 16, 1932, Rockefeller Foundation Archives, North Tarrytown, New York, 1.1./series 717 A, box 10, folder 63. See also: Harald Kranz, "Rassenhygiene/Eugenik in Deutschland: Institutionalisierung und Politisierung einer Wissenschaft (1927–1945)," thesis, University of Bielefeld, 1984, 55–57.

33. Human Heredity Committee of the International Federation of Eugenic Organization, Davenport Papers: IFEO. The other studies mentioned were by Swedish eugenicists Hermann Lundborg and Torsten Sjögren on idiocy and oligophrenia, and by Dutch eugenicist Waardenburg on the heredity of eye diseases.

34. Davenport Papers: Fischer. See: Marvin D. Miller, *Wunderlich's Salute: The Interrelationship of the German–American Bund, Camp Siegfried, Yaphank, Long Island, and the Young Siegfrieds and their Relationship with American and Nazi Institutions* (New York: Malamud-Rose, 1983).

35. Minutes of the Conference, Davenport Papers: Gini.

36. Minutes of the Committee on Race Psychiatry in Rome, 1929, Davenport Papers: IFEO Rom 1929.

37. Fischer to Davenport, January 1, 1932, Davenport Papers: Fischer.

38. Davenport Papers: International Congress of Eugenics. The Associated Press reported that it "regards this congress as one of the most interesting scientific news events of the year."

39. *Eugenic News*, 18 (1933):14.

40. *Eugenic News*, 18 (1933): 14. See: Weindling, *Health*, 452–53.

41. Heinrich Schopohl, "Die Eugenik im Dienste der Volkswohlfahrt," *Volkswohlfahrt*, 13 (1932): 469–587.

42. See: Joachim Müller, *Sterilisation und Gesetzgebung bis 1933* (Husum: Abhandlungen zur Geschichte der Medizin und der Naturwissenschaft, 1985): 60–61; Weingart, Kroll, Bayertz, *Rasse*, 291–292.

43. Gerhard Boeters, "Die Unfruchtbarmachung der geistig Minderwertigen," *Sächsische Staatszeitung*, July 9–11, 1923, quoted in Proctor, *Racial Hygiene*, 98.

44. Zwickau to the Auswärtige Amt, October 8, 1923, BAP, RMI 9347.

45. Bumm to the Reich Ministry of the Interior, October 15, 1923, BAP, RMI 9247.

46. Bumm to the Reich Ministry of the Interior, January 23, 1924, BAP, RMI 9247.

47. Professor Otto Reche, chairman of the Vienna Society for Racial Care (Gesellschaft für Rassenpflege), expressed his concern over the fact that the Americans had become the world leaders in racial hygiene and urged Germans to catch up. Otto Reche, "Die Bedeutung der Rassenpflege für die Zukunft unseres Volkes," *Veröffentlichungen der Wiener Gesellschaft für Rassenpflege*, 1 (1925): 6.

48. Nowak, *"Euthanasie" und Sterilisierung*, 70.

49. For information concerning the interest of Social Democrats in eugenics, see: Weingart, Kroll, Bayertz, *Rasse*, 105–14 and Michael Schwartz, "Sozialistische Eugenik: Eugenische Sozialtechnologien in Diskurs und Politik der deutschen Sozialdemokratie, 1890–1933," diss., University Münster, 1992.

50. Robert Gaupp, *Die Unfruchtbarmachung geistig und sittlich Kranker und Minderwertiger* (Berlin: Julius Springer, 1925): 8–9. Philip R. Reilly, *The Surgical Solution: A History of Involuntary Sterilization in the United States* (Baltimore: Johns Hopkins University Press, 1991): 97.

51. Gaupp, *Unfruchtbarmachung*, 42.

52. Harry H. Laughlin, "Die Entwicklung der gesetzlichen rassenhygienischen Sterilisierung in den Vereinigten Staaten," *Archiv für Rassen- und Gesellschaftsbiologie*, 21 (1929): 253–62; quotes are on page 262.

53. Eugene S. Gosney and Paul Popenoe, *Sterilization for Human Betterment* (New York: Milliam, 1929). The slightly abridged German translation appeared under

the title: *Sterilisierung zum Zwecke der Aufbesserung des Menschengeschlechts,* trans. Konrad Burchardi (Berlin: Marcus & Webers, 1930).

54. Felix Tietze, "Sterilisierung zum Zwecke der Aufbesserung des Menschengeschlechts," *Archiv für Rassen- und Gesellschaftsbiologie,* 25 (1931): 346–47. See also the reference in Jenny Blasbalg, "Ausländische und deutsche Gesetzentwürfe über Unfruchtbarmachung," *Zeitschrift für die gesamte Strafrechtswissenschaft,* 52 (1932): 477–96.

55. Otto Kankeleit, *Die Unfruchtbarmachung aus rassenhygienischen und sozialen Gründen* (Munich: Lehmann, 1929): 95. Laughlin, *Entwicklung,* 260–61. Similar arguments were advanced by Protestant sterilization expert Hans Harmsen in a meeting of the eugenic committee of the Protestant Church (Ständiger Ausschuss für Fragen der Rassenhygiene und Rassenpflege beim Central-Ausschuss für die Innere Mission der Deutschen Evangelischen Kirche) in May 1931, Archiv des Diakonischen Werkes, Berlin (abbr. ADW), CA/G 1800/1. See: Bock, *Zwangssterilisation,* 373. Articles by Otto Kankeleit include: "Die Ausschaltung geistig Minderwertiger von der Fortpflanzung," *Volk und Rasse,* 6 (1931): 174–79; "Künstliche Unfruchtbarmachung aus rassenhygienischen und sozialen Gründen," *Zeitschrift für die gesamte Neurologie und Psychiatrie,* 98 (1925): 220–53.

56. Frithjof Hager, "Der gegenwärtige Stand der Frage der Sterilisierung Minderwertiger in Deutschland," diss., University of Kiel, 1934, 16; Helmut Kuhlberg, "Die Auswirkung des Gesetzes zur Verhütung erbkranken Nachwuchses in der Heil- und Pflegeanstalt Waldbröl," diss., University Bonn, 1934, 5; Dora Neeff, "Die bisherigen Erfahrungen über Eingriff und Verlauf der sterilisierenden Operation bei der Frau," diss., University of Heidelberg, 1935, 5; Theo Osterfeld, "Über die Sterilisation aus eugenischer Indikation," diss., University of Würzburg, 1936, 6; Werner Bauer, "Erste Erfahrungen mit der Anwendung des Sterilisierungsgesetzes bei Geisteskranken," diss., University of Tübingen, 1936, 3–5; Walter Kreienberg, "Die Auswirkungen des Gesetzes zur Verhütung erbkranken Nachwuchses an dem Krankenbestand der Psychiatrischen und Nervenklinik Erlangen," diss., University of Erlangen, 1937, 3; Otto Striehn, "Kastration," diss., University of Munich, 1938, 15; Erika Holzmann, "Erfahrungen und Ergebnisse der Untersuchungen auf Ehetauglichkeit in Hamburg vom 20. Oktober 1935 bis 1. Juli 1940," diss., University of Rostock, 1941, 6–8. I am indebted to Gisela Bock for providing me access to several medical dissertations written between 1933 and 1945.

57. Herbert Hüllstrung, "Über gesetzliche Bestimmungen und Erfolge der Zwangssterilisierung und Zwangskastration," diss., University of Bonn, 1934, 3.

58. W. Knapp, "Statistisches und Empirisches über das Gesetz zur Verhütung erbkranken Nachwuchses," diss., University of Bonn, 1934, 6–7.

59. Ursula Krause, "Erfahrungen und Ergebnisse bei 315 Sterilisationen aus eugenischer Indikation vom 4. April 1934 bis zum 1. April 1936," diss., University of Kiel, 1937, 5–6.

60. Walter Schultze, "Der nordische Gedanke," *Ziel und Weg,* 2 (1932): 4. See also: Robert Proctor, "Nazi Biomedical Technologies," *Lifeworld and Technology,* eds. Timothy Casey and Lester Embree (Washington, D.C.: The Center for Advanced Research in Phenomenology and University Press of America, 1989): 33.

61. Adolf Hitler, *Mein Kampf,* 1924; trans., *Mein Kampf,* with an introduction by D. C. Watt (London 1974): 400.

Chapter 3

1. *Time* (September 9, 1935): 20–21, quoted in: Kevles, *Name*, 347.
2. International Federation of Eugenic Organizations, Eleventh Conference, Zurich 1934, *Bericht über die 11. Versammlung der Internationalen Föderation eugenischer Organisationen, Konferenzsitzungen vom 18. bis 21. Juli 1934 im Waldhaus Dolden, Zurich* (Zurich, O. Füssli, 1934): 79.
3. Hodson, secretary of the International Federation, from a report in 1934, Davenport papers: IFEO. Max Weinreich, *Hitler's Professors: The Part of Scholarship in Germany's Crimes against the Jewish People* (New York: Yiddish Scientific Institute, 1946).
4. The resolution is printed in: Heinz Kürten, "Der Kongress der Internationalen Föderation Eugenischer Organisationen," *Ziel und Weg*, 4 (1934): 15–16. See also: *Volk und Rasse* (1934): 298–99 and Rudolf Frercks, *Deutsche Rassenpolitik* (Leipzig: Reclam, 1937): 53.
5. His speech was printed in a special edition of the *Rassenpoltische Auslandskorrespondenz* (1935), no. 3. See also: *Neues Volk*, 2.9 (1934): 13.
6. For most of his life, Rüdin was a Swiss citizen. He accepted German nationality relatively late. See: Bock, *Zwangssterilisation*, 25, 484.
7. The exclusion of Jews, couples with mental or physical illnesses, and political opponents was part of a decree from June 6, 1933.
8. International Federation of Eugenic Organizations, *Bericht der 12. Versammlung der Internationalen Federation Eugenischer Organisationen, Konferenzsitzungen vom 15. bis 20. Juli 1936, Scheveningen, Holland* (Den Haag: W. P. van Stockum, 1936): 56–65 (my translations). His speech was also printed in a German journal: Falk Ruttke, "Erbpflege in der Deutschen Gesetzgebung," *Ziel und Weg*, 4 (1934): 600–3. For information on the unification of the Nazi health administration, see: Alfons Labisch and Florian Tennstedt, *Der Weg Zum "Gesetz über die Vereinheitlichung des Gesundheitswesens" vom 3. Juli 1934* (Düsseldorf: Akademie für öffentliches Gesundheitswesen, 1985).
9. *Journal of Heredity*, 26 (1935): 10.
10. "Organizations: The Twelfth Meeting of the International Federation of Eugenic Organizations," *Eugenic News*, 21 (1936): 106–14. The quote about Goethe's talk is from a report by Dutch eugenicist Jacob Sanders, p. 106.
11. Minutes of the 1936 Scheveningen Meeting, Davenport Papers: IFEO. See also: International Federation of Eugenic Organization, *Bericht*.
12. Abortions for "hereditarily valuable" women were only allowed for medical reasons. See: Bock, *Zwangssterilisation*, 98–99, 383–84.
13. Falk Ruttke, "Erbpflege in der deutschen Gesetzgebung," *Der Erbarzt*, 3 (1936): 111–17. Quote is my translation from p. 117.
14. See, for example, the reports in: *Volk und Rasse*, 8 (1936): 288; *Rassenpolitische Auslandskorrespondenz*, 3.7/8 (1936): 11; *Der Erbarzt*, 3 (1936): 117, 174–75; "Nationalsozialistische Rassenforschung im Urteil des Auslands," *Westfälische Zeitung* (August 10, 1936); the same article was printed under the title "Deutschlands eugenische Maßnahmen im Urteil des Auslandes," *Völkischer Beobachter* (August 11, 1936).
15. "Deutschland führend in der Vererbungsforschung," *Völkischer Beobachter*, August 2, 1936 (my translations). See also *Volk und Rasse*, 11 (1936): 436.

16. Letter from Rodenwaldt; March 30, 1936, Bundesarchiv Potsdam (BAP, REM 4901–3171, 5), (my translation). The other institutes were the newly founded Institute for Hereditary Biology and Racial Hygiene in Frankfurt and the Institute for Education of the University of Thuringia.
17. Letter to Thummala on February 16, 1936, BAP, REM 4901–3198.
18. The IFEO agreed to a conference in Germany on January 9, 1939 (BAP, REM 4901–2839, 140). The international announcement of the conference appeared in *Rassenpolitische Auslandskorrespondenz*, 6.7 (1939): 5. Copies of the official invitation in four languages can be found at the Bayerisches Hauptstaatsarchiv, Munich, M Inn 79477 Rassenhygiene and in BAK R18/3592, p. 130. Because of the international political situation, the conference never took place. See Herbert Linden from the Department of the Interior to the Department for Education on October 11, 1940 (BAP, REM 4901–2839, p. 153).
19. For the shift of Raymond Pearl from a eugenicist to an early population scientist who preserved a eugenic ideology, see: Garland E. Allen, "Old Wine in New Bottles: From Eugenics to Population Control in the Work of Raymond Pearl," *The Expansion of American Biology*, eds. Keith R. Benson, Jane Maienschein and Ronald Rainger (New Brunswick, London: Rutgers University Press, 1991): 231–61.
20. Pearl to Sir Charles Close, British eugenicist and population scientist and leading member of the IUSIPP, December 17, 1934, Pearl Papers, APS: IUSIPP.
21. Pearl to Fischer, April 1935, and Pearl to Close, January 16, 1935. For his critical reaction to the results of the conference, see: Pearl to Close, February 11, 1936, Pearl Papers.
22. Harry H. Laughlin, "Studies on the historical and legal development of eugenic sterilization in the United States," *Bevölkerungsfragen: Bericht des Internationalen Kongresses für Bevölkerungswissenschaft Berlin 26.August–1.September 1935*, eds. Hans Harmsen and Franz Lohse (Munich: Lehmann, 1936): 664–75.
23. Laughlin to Fischer, July 31, 1935, Laughlin Papers. See: Hassencahl, *Laughlin*, 341. Hassencahl remarks that Laughlin's paper for the Berlin conference was essentially a version of his pamphlet, "The Legal Status of Sterilization," published as *Supplement to the Annual Report of the Municipal Court of Chicago for the Year 1929*.
24. Clarence G. Campbell, "The Biological Postulates of Population Study," *Bevölkerungsfragen: Bericht des Internationalen Kongresses für Bevölkerungswissenschaft Berlin 26.August–1.September 1935*, eds. Hans Harmsen and Franz Lohse (Munich: Lehmann, 1936): 602.
25. Eugen Fischer, "Schlussansprache," *Bevölkerungsfragen: Bericht des Internationalen Kongresses für Bevölkerungswissenschaft Berlin 26.August–1.September 1935*, ed. Hans Harmsen and Franz Lohse (Munich: Lehmann, 1936): 928 (my translation).
26. Heinrich Schade, "Der Internationale Kongress für Bevölkerungswissenschaft in Berlin," *Der Erbarzt* 2 (1935): 140–41. For commentaries in the public press, see, for example: "Internationale Kongresse in Berlin: Die Bevoelkerungswissenschaftler der Welt lernen die nationalsozialistische Erb- und Rassenpflege kennen," *Völkischer Beobachter*, August 25, 1935; "Ausländische Anerkennung für das Thüringische Landesamt für Rassenwesen in Weimar," *Allgemeine Thüringische Landeszeitung*, September 8, 1935. This article gave an

account of a telegram full of appreciation from the participants of the Congress to the German Secretary of the Interior and honorary president of the Congress, Wilhelm Frick.

27. "Deutsche Wissenschaftler machen jetzt Weltgeschichte—Ein Amerikaner über rassische Pflichten," *Völkischer Beobachter,* August 29, 1935 (my translation).

28. "Deutschlands rassenpolitische Massnahmen Vorbild für das Ausland," *Berliner Börsenzeitung,* September 5, 1935.

29. Clarence G. Campbell, "The German Racial Policy," *Eugenic News,* 21 (1936): 25.

30. Campbell, *German Racial Policy,* 28, 26.

31. Clarence C. Campbell, "Die deutsche Rassenpolitik," trans. Inge Bass, *Der Erbarzt* 3 (1936): 140–42. The journal of the Propaganda Office for Population Policy and Racial Welfare for German foreign scientists and politicians printed a similar article: Clarence C. Campbell, "Urteil über die deutsche Rassenpolitik," *Rassenpolitische Auslandskorrespondenz* 3.10 (1936): 2–3.

32. "Sterilisationsgesetz im Urteil der Welt," *Angriff* (July 10, 1936) and "Um die Zukunft der Völker: Die rassenhygienische Besinnung macht Fortschritte—Auflehnung der amerikanischen Eugeniker gegen die zunehmende Degeneration," *NSK* (November 12, 1936). See also: E. Hester, "Deutschlands Rassenpolitik im Urteil der anderen: Amerikanische und englische Anerkennung für die deutsche Rassenpolitik," *Völkischer Beobachter* (August 19, 1936) and E. Hester, "Ausländer über deutsche Rassenpolitik," *Preussische Zeitung* (August 31, 1936).

33. "Amerikanische Forscher fordern Anwendung des Sterilisationsgesetzes in der ganzen Welt," *NSK* (March 11, 1936). For general information about Campbell's speech in Toronto, see: McLaren, *Master Race,* 122.

34. For the development of the multiple gene theory, see: Kenneth M. Ludmerer, "American Geneticists and the Eugenics Movement 1905–1935," *Journal of the History of Biology* 2 (1969): 32. For the influence of Morgan's Chromosome Theory in Germany, see: Jonathan Harwood, "The Reception of Morgan's Chromosome Theory in Germany: Interwar Debate over Cytoplasmic Inheritance," *Medizinhistorisches Journal,* 19 (1984): 3–32. For the shift in the scientific position on race crossing, see: William Provine, "Geneticists and the Biology of Race Crossing," *Science,* 182 (1973): 790–96.

35. Leon F. Whitney, *The Case for Sterilization* (New York: Frederick A. Stocks, 1934): 7.

36. Report of the Sub-Committee on the Ultimate Program for the Eugenics Society of American, Introduction, Jennings Papers, APS: Eugenics Society of America.

37. See Herbert S. Jennings, *The Biological Basis of Human Nature* (New York: W. W. Norton, 1930): 203.

38. Falk Ruttke, "Erbpflege in der deutschen Gesetzgebung," *Der Erbarzt* 3 (1936): 113. See also: International Federation of Eugenic Organization, *Bericht.*

39. Rudolf Hess at a mass meeting in 1934. See: Robert Lifton, *The Nazi Doctors: Medical Killing and the Psychology of Genocide* (New York: Basic Books, 1986): 31. Fritz Lenz, *Menschliche Auslese und Rassenhygiene (Eugenik),* (Munich: Lehmann, third edition 1931): 417. This popular book was part of a new edition of the Baur, Fischer, Lenz. I thank Robert Proctor for informing me of the intellectual origin of the Hess quote. See: Proctor, *Nazi Biomedical Technologies,* 23.

40. *Eugenic News* 20 (1935): 100.
41. Paul Popenoe, "The German Sterilization Law," *Journal of Heredity*, 25 (1934): 257. See also: Mehler, *Elimination*, 15.

Chapter 4

1. Richmond *Times-Dispatch*, February 27, 1980; March 2, 1980. Quoted in Kevles, *Name*, 116.
2. Ten articles in the *Rassenpolitische Auslandskorrespondenz* on Great Britain dealt mostly with Britain's attempts to establish a voluntary sterilization law. France was only important because of its attempt to raise the birth rate. No article treated the position of Italy toward eugenic racism. Reports were limited to nationalism in Italy. A few notes mentioned the implementation of race policy in Scandinavian countries.
3. Otto Wagener, *Hitler aus nächster Nähe: Aufzeichnungen eines Vertrauten 1929–1932*, ed. Henry A. Turner (Frankfurt a.M.: Ullstein, 1978): 264. Quoted in Trombley, *Right*, 116. About Otto Wagener's importance as confidante of Hitler, see: Erich Czech-Jochberg, *Adolf Hitler und sein Stab* (Oldenburg: Gerd Stalling, 1933): 109–11. The best secondary literature about Wagener seems to be: Henry A. Turner, *Die Großunternehmer und der Aufstieg Hitlers* (Berlin: Siedler, 1985). For the use of Wagener's memories as historical source see the introduction of Turner to *Hitler aus nächster Nähe*, i–xvii.
4. "Das Ausland als Vorbild für die deutsche Rassengesetzgebung," *Rassenpolitische Auslandskorrespondenz*, 2.4 (1935): 1
5. Jens Paulsen, "Politische Entwicklungsrichtungen in den europäischen Völkern," *Archiv für Rassen und Gesellschaftsbiologie*, 33 (1939): 224.
6. *Volk und Rasse*, 9 (1934): 398. The statement from Supreme Court Judge Oliver Wendell Holmes is quoted in: Ellsworth Huntington, *Tomorrow's Children: The Goal of Eugenics* (New York: John Wiley, 1935): 43.
7. Paul Heinz Besselmann, "Nationale eugenische Ausleseprobleme in der Volkswirtschaftslehre," diss., University Heidelberg, 1934, 67. On sterilization in the United States: Bruno Steinwallner, "Rassenhygienische Gesetzgebung und Maßnahmen ausmerzender Art," *Fortschritte der Erbpathologie, Rassenhygiene und ihrer Grenzgebiete*, 1 (1937): 203–5; Reimer Schulz, "Das Gesetz zur Verhütung erbkranken Nachwuchses im Spiegel amerikanischer und deutsch-amerikanischer Presse," *Schleswig-Holsteinische Hochschulblätter*, 34 (1934): 19–20.
8. H. H. von Schneidewind, *Wirtschaft und Wirtschaftspolitik der Vereinigten Staaten von Amerika* (Würzburg: Konrad Triltsch, 1933): 29–32. Although this study was finished in 1928, it is revealing that it was only prepared for publication shortly after the takeover by the Nazis in January 1933.
9. Günther at a meeting of the "Kampfbundes für deutsche Kultur" on February 21, 1934. "Der Vererbungs- und Rassegedanke innerhalb der Einwanderungsgesetzgebung," *Völkischer Beobachter* (February 23, 1934).
10. Steinwallner, 249–50. Walter Gross, "Die Welt und der Rassengedanke in Deutschland," *Zeitungsdienst* (November 2, 1934).
11. Laughlin's Model Law is printed in: Harry H. Laughlin, *Eugenic Sterilization in the United States* (Chicago: Psychopathic Laboratory of the Municipal Court of

Chicago, 1922): 446–47. See also: Johnson, *International Eugenics*, 120; Chase, *Legacy*, 449; Lapon, *Mass Murder*, 16; Smith, *Minds Made Feeble*, 154.

12. *Eugenic News*, 18 (1933): 89. On the influence of the German sterilization law: Johnson, "International Eugenics," 120, and Marie E. Kopp, "A Eugenic Program in Operation," *Summary of the Proceeding of the Conference on Eugenics in Relation to Nursing*, ed. Frederick Osborn, unpublished, American Eugenics Society Papers, APS, Conference on Eugenics in Relation to Nursing. A German dissertation from the 1930s contradicted this view: Paul Wappler, "Über die gesetzliche Sterilisation und unsere Erfahrungen hiermit an Hand von 220 Sterilisierungen," diss., University of Leipzig, 1937, 3.

13. Sauckel to the Reich Ministry of the Interior, April 4, 1933, BAK R 43 II/717. See also: Bock, *Zwangssterilisation*, 82.

14. Ruttke, *Erbpflege*, 601.

15. For complaints that legislators did not learn enough from the experiences in the United States, see: Jacob H. Landman, *Human Sterilization: The History of the Sexual Sterilization Movement* (New York: Macmillan, 1932), 218.

16. Henry Herbert Goddard, *The Kallikak Family: A Study in the Heredity of Feeble-Mindedness* (New York: Macmillan, 1912); *Feeble-Mindedness: Its Causes and Consequences* (New York: Macmillan, 1914). See: Smith, *Minds Made Feeble*. Goddard is quoted in: Chase, *Legacy*, 151–52.

17. Smith, *Minds Made Feeble*, 139–49.

18. Quoted in: Chase, *Legacy*, 153.

19. Karl Winckler (trans.), "Die Familie Kallikak," *Friedrich Mann's Pädagogisches Magazin*, No. 1393 (1934): ii. Quoted in: Smith, *Minds Made Feeble*, 161–162. See also: Hans Krauss, *Die Grundgedanken der Erbkunde und Rassenhygiene in Frage und Antwort* (Munich: Otto Gmelin, 1935): 50–51.

20. *Neues Volk*, 2 (1934): 47 (my translation).

21. "Wer ist die Familie Kallikak," *Wochenblatt Sachsen Anhalt* (October 24, 1935) (my translation). See also: *Volk und Rasse*, 9 (1937): 387–388.

22. See: Henry Herbert Goddard, "Feeblemindedness," *Journal of Psycho-Asthenics*, 33 (1928): 219–27. See also: Bock, *Zwangssterilisation*, 115 and 343.

23. "Ein Verbrecher—2820 Nachkommen," *Z.d.R.* (March 27, 1937).

24. Marie E. Kopp, "Legal and Medical Aspects of Eugenic Sterilization in Germany," *American Sociological Review*, 1 (1936): 763. A similar statement appears in: Marie E. Kopp, "The German Sterilization Program," undated typescript, Association for Voluntary Sterilization Papers, University of Minnesota, Minneapolis. See: Reilly, *Surgical Solution*, 106.

25. Gosney and Popenoe, *Sterilization*. An article by Popenoe published in the main German eugenic journal was based on this study: Paul Popenoe, "Rassenhygienische Sterilisierung in Kalifornien," *Archiv für Rassen- und Gesellschaftsbiologie*, 23 (1931): 249–59.

26. "Sterilization in Kalifornien: Ein segenreiches Gesetz, welches seit 26 Jahren in Kraft ist," *Völkischer Beobachter* (December 19, 1935). See also: "Sterilisierung zum Zweck der Aufbesserung des Menschengeschlechts," *Gesundheitsführung* (1934): 60.

27. Human Betterment Foundation, *Human Sterilization*, (Pasadena, Calif.: Human Betterment Foundation, n.d.). Part of the pamphlet is reprinted in: "Sterilisierung

in den Vereinigten Staaten," *Rassenpolitische Auslandskorrespondenz*, 1.1 (1934): 4.

28. Gosney to Wally Schick, October 20, 1933. ADW, CA/GF 2000, I4. Hans Harmsen before the Arbeitsgemeinschaft für Volksgesundung. ADW, CA/GF 2000, I1 6aff. For excellent studies of the position of the Inneren Mission toward eugenics, see: Kaiser, *Innere Mission* and *Sozialer Protestantismus*.

29. Niederschrift über die Beratung des Sachverständigen Beirates für Bevölkerungs- und Rassenpolitik, June 25, 1934, BAP 15.01 RMI 26229. Gütt received the pamphlet from the NSDAP Reichsleitung on February 7, 1934.

30. *Der Erbarzt*, 3 (1936): 175. The article was reprinted under the title "Deutsches Vorbild macht Schule: Rassenhygiene im Ausland," *Fränkische Tageszeitung*, (April 3–4, 1937).

31. "Umfrage über die Sterilisation," *NSK*, July 28, 1936 (my translation). See also a note in: *Ziel und Weg*, 6 (1936): 112.

32. The basis for the information was provided by Paul Popenoe and Eugene S. Gosney, *Twenty-eight Years of Sterilization in California* (Pasadena, Calif.: Human Betterment Foundation, 1939), and Human Betterment Foundation, *Human Sterilization Today*, (Pasadena, Calif.: Human Betterment Foundation, 1939). See also: note in *Zeitschrift für Rassenkunde*, 2 (1936): 224, about the twenty-fifth anniversary of the sterilization law in California. "Sterilisierung in den Vereinigten Staaten," *Völkischer Beobachter*, August 14, 1938. Note in: *Archiv für Rassen- und Gesellschaftsbiologie*, 33 (1939): 448.

33. Quoted according to Trombley, *Right*, 117, from which I took information about the exhibition. It was not possible for me to find more information about the German organizer of the exhibition who met eugenicists in Pasadena, Los Angeles, New York, Chicago, and Baltimore during his tour through the United States.

34. Popenoe, *German*, 257.

35. Paul Popenoe, "The Progress of Eugenic Sterilization," *Journal of Heredity*, 25 (1934): 19–25.

36. Popenoe to L. C. Dunn, January 22, 1934, Dunn Papers, APS: Popenoe, quoted in Ludmerer, *Genetics*, 118.

37. Paul Popenoe, "Trends in Human Sterilization," *Eugenic News*, 22 (1937): 42–43. Ernst Rüdin, "Strömungen in Art und Umfang der Sterilisationspraxis," *Archiv für Rassen- und Gesellschaftsbiologie*, 31 (1937): 367–69. Rüdin expressed the opinion that the German sterilization law fulfilled all the demands of Popenoe's evaluation.

38. Kevles, *Name*, 116.

39. Information according to Trombley, *Right*, 115.

40. The Sterilization League of New Jersey was the basis for the national organization Birthright, which later became the Association for Voluntary Sterilization. When the Sterilization League of New Jersey changed its name to "Birthright" in 1943, Marion S. Norton changed her name to Marian S. Olden. She changed her first name to be clearly identified as a woman. In 1970 she published a trilogy about the history of eugenics: Marian S. Olden, *Human Betterment Was Our Goal* (Gwynedd, 1970); *Sterilization League of New Jersey, 1937–1942* (Gwynedd, 1970); *From Birthright, Inc., to Voluntary Sterilization* (Gwynedd, 1970).

41. Davenport Papers, Leon F. Whitney, "Memoirs," 204; Kevles, *Name*, 118.

42. Whitney, *Case*, 137.
43. *Eugenic News*, 19 (1934): 102.
44. *Eugenic News*, 18 (1933): 89–90.
45. C. Thomalia, "The Sterilization Law in Germany," trans. Alice Hellmer, *Eugenic News*, 19 (1934): 137–40.
46. Translation of the German sterilization law in *Eugenic News*, 18 (1933): 91–93. Frick's speech before the first meeting of the Expert Council for Population and Race Politics, held in Berlin on June 28, 1933, is reprinted in *Eugenic News*, 19 (1934): 33–34. Best wishes for Verschuer's institute appeared in *Eugenic News*, 21 (1936): 59–60. The article "Jewish Physicians in Berlin" was printed in *Eugenic News*, 19 (1934): 126.
47. The article on the positive reactions of *Eugenic News* was printed in *Rassenpolitische Auslandskorrespondenz*, 2.7 (1935): 6. A similar article appeared in *Volk und Rasse*, 10 (1935): 155–56. The *Zeitschrift für Rassenkunde* reported on several articles of *Eugenic News* in 1936, including a report on Native Americans, the founding of a "Society for Constructive Eugenics in Philadelphia," and the relationship of intelligence quotients and eugenics. The correction of the figures from the *Eugenic News*, 20.6 (1935) appeared in *Volk und Rasse*, 11 (1936): 152.
48. The article "Nazis Open Race Bureau for Eugenic Segregation" was from *The New York Times* (May 4, 1933). The expression about Hitler is in the Laughlin Papers, "Hitler and the Jews," Clippings, 1933–1934. Historians are not sure if the handwriting is from Laughlin. Jerry Hirsch and Thomas M. Shapiro have identified the handwriting as Laughlin's. Shapiro, *Population*, 206. Proctor claims that it belongs to another anonymous member of the ERA. Proctor, *Racial Hygiene*, 101.
49. Martin S. Pernick, University of Michigan Historical Health Film Collection, has informed me that a German version of the movie exists in the National Archives Film Division in Washington, D.C., reel number 243.3.
50. *Informationsdienst des Rassenpolitischen Amtes der NSDAP*, 21 (August 30, 1937): section 0802. See also a statement of the Rassenpolitischen Amtes from March 3, 1936 that Hitler saw *Erbkrank* on February 26, 1936. The information from the censorship office is in: Bundesarchiv Koblenz, Filmarchiv: Entscheidungen der Filmprüfstelle. See also: Report about *Erbkrank* in *Völkischer Beobachter* (May 6, 1937): 6. All information about the German reception of *Erbkrank* is from Karl Ludwig Rost's comprehensive book, *Sterilisation und Euthanasie im Film des "Dritten Reiches"* (Husum: Matthiesen, 1987). See also: Michael Burleigh, " 'Euthanasia' and the Cinema in Nazi Germany," *History Today*, 40 (1990): 11–16; James M. Gardener, "Contribution of the German Cinema to the Nazi Euthanasia Program," *Mental Retardation*, 20 (1982): 174–75; Martin S. Pernick, *The Black Stork: Eugenics and the Death of "Defective" Babies in American Medicine and Motion Pictures since 1915* (New York: Oxford University Press, 1992).
51. See: Rost, *Sterilisation*, 226 (my translation).
52. See: Rost, *Sterilisation*, 226–28 (my translation).
53. Harry H. Laughlin, "Eugenics in Germany: Motion Picture Showing how Germany Is Presenting and Attacking Her Problems in Applied Eugenics," *Eugenic News*, 22 (1937): 65–66. One subtitle states, for example, that "the Jewish folk

provides a particularly high percentage of feebleminded people." See: Rost, *Sterilisation,* 227 (my translation).

54. BAK, NS 2/85, fol. 280, see: Proctor, *Racial Hygiene,* 101.

55. Letter from Laughlin to W. P. Draper, December 9, 1938, Laughlin Papers. Draper financed Davenport's research on race crossing in Jamaica, the translation of eugenics texts from German into English, and provided money for the publication of his essays on comparative birth rates. Davenport Papers: Draper. See: Hassencahl, *Laughlin,* 355–356 and Mehler, *Foundation,* 21.

56. "Rassenpolitische Aufklärung nach deutschem Vorbild: Grosse Beachtung durch die amerikanische Wissenschaft," *NSK,* March 19, 1937, and notes in *Ziel und Weg,* 7 (1937): 361, and in *Volk und Rasse,* 12 (1937): 150.

57. Kopp, *Eugenic Program,* 4.

58. Whitney, *Case,* 136.

59. "Eugenic Sterilization in Germany," *Eugenic News,* 88 (1933): 89.

60. Report about an article of the Reich minister of justice, Franz Günther, about "Das neue Reichsgesetz gegen gefährliche Gewohnheitsverbrecher (The New Law against Dangerous Habitual Criminals)," *Archiv für Kriminologie,* 93 (1933): 197–200. See: *Eugenic News,* 19 (1934): 79.

61. See, for example: "Welche Staaten bekämpfen Erbkrankheiten," *Völkischer Beobachter* (March 9, 1938). "Geheime zwangsweise Sterilisation von Verbrechern in den USA," *Deutsches Nachrichtenbüro* (May 3, 1942).

62. It should be mentioned that after 1937 sterilization was applied to religious minorities as well as to ethnic minorities. For example descendents of German women and black French soldiers during the occupation of the Ruhrgebiet after World War I, as well as Gypsies, were sterilized. For the sterilization of Jews, Gypsies, blacks, and Poles, see: Bock, *Zwangssterilisation,* 351–68.

63. Whitney, *Case,* 138.

64. Robert Cook, "A Year of German Sterilization," *Journal of Heredity,* 26 (1935): 485–89; Heinrich Schade, "Ausländische Stimmen zur deutschen Erb- und Rassenpflege," *Rassenpolitische Auslandskorrespondenz,* 3.5 (1936): 3–4.

65. Quoted in: Paul Weindling, "Race, Blood, and Politics," *Times Higher Education* (July 19, 1985): 19.

Chapter 5

1. Lothrop Stoddard, *Into the Darkness: Nazi Germany Today* (New York: Deull, Sloan & Pearce, 1940): 190.

2. Whitney, *Case,* 7

3. William W. Peter, "German Sterilization Program," *American Journal of Public Health,* 24 (1934): 187–91.

4. Peter, "German," 189–90.

5. William W. Peter, "Ein amerikanischer Arzt zur Sterilisierung in Deutschland," *Neues Volk,* 2 (1934): 16–18 (my translations).

6. Marie E. Kopp before the Conference of the American Eugenics Society on the Relation of Eugenics to Nursing, 1937. Kopp, Eugenic Program, 1. She also published: "Eugenic Sterilization Laws in Europe," *American Journal of Obstetrics and Gynecology,* 34 (1937): 499; "Eugenic Sterilization Laws in Europe," *Bulletin of the New York Academy of Medicine,* 34 (1937): 761–69; "Surgical

Treatment of Sex Crime Prevention Measure," *Journal of Criminal Law, Criminology and Police Science*, 28 (1938): 692–706.

7. Marie Kopp at the 25th Annual Meeting of the Eugenics Research Association, New York, June 5, 1937, Abstract, printed in *Eugenic News*, 22 (1937): 60.

8. Quoted in: Olden, *Human Betterment*, 72.

9. Kopp, Eugenic Program, 3. Kopp, "Legal and Medical Aspects," 767. Bock, *Zwangssterilisation*, 380, had documented that because of the more serious operation for women, 90 percent of the 5,000 people who died as a result of sterilization procedures were women. These people should be regarded as the first victims of the mass killings of Nazi Germany.

10. Kopp, Eugenic Program, 4. Summary of the Proceedings of the Conference on Eugenics in Relation to Nursing, February 24, 1937 by Frederick Osborn, American Eugenics Society Paper, APS. See also: Marie E. Kopp, "The German Program of Marriage Promotion through State Loan," *Eugenic News*, 21 (1936): 121–29.

11. Charles M. Goethe is neglected in the historiography of eugenics. The information about Goethe is from Olden, *Sterilization Leaque*, 143; Mehler, *History*, 355–56; Reilly, *Surgical Solution*, 67. Publications by Goethe during his presidency of the Eugenics Research Association include: Charles M. Goethe, "Patriotism and Racial Standards," *Eugenic News*, 21 (1936): 65–69; "Eugenics and Geography," *Eugenic News*, 22 (1937): 47; "Extinction of the Inca Highcastes: Presidential Address before the Twenty-Fifth Annual Meeting of the Eugenics Research Association," *Eugenic News*, 22 (1937): 51–57.

12. Paul Popenoe reprinted the letter in the annual report of the Human Betterment Foundation for 1935. See: Trombley, *Right*, 117.

13. Marion S. Norton, *Selective Sterilization in Primer Form* (Princeton: Sterilization Leaque of New Jersey, 1937). Quote in: Olden, *Sterilization League*, 139.

14. Letter "Der Reichs- und Preussische Minister des Innern" to "Chef der Reichskanzlei," September 18, 1937, BAK, R 43 II, 721a. translation of Norton's pamphlet was in the files of the Reich Minister of the Interior, BAK, RMI R 18, 5585. About the reaction of the German Catholic Church to the sterilization laws see: Nowak, *Euthanasie*; Bock, *Zwangssterilisation*, 289–98.

15. Olden, *Sterilization Leaque*, 144. Marion S. Norton, *Major Provisions for Population Control Abroad* (Princeton: Sterilization Leaque of New Jersey, 1939).

16. Weindling, *Health*, 556. Benno Müller-Hill, *Die Tödliche Wissenschaft: Die Aussonderung von Juden, Zigeunern und Geisteskranken 1933–1945* (Reinbek by Hamburg: Rowohlt, 1984): 108–9.

17. T. U. H. Ellinger, "On the Breeding of Aryans and Other Genetic Problems of War-Time Germany," *Journal of Heredity*, 33 (1942): 141–43.

18. Richard Goldschmidt, "Anthropological Determination of 'Aryanism,'" *Journal of Heredity*, 33 (1942): 215–16.

19. Goldschmidt openly favored a sterilization law and did not realize before 1933 that anti-Semitism could become institutionalized in the apparatus of eugenics and hereditary research, which he helped to establish. Weindling, *Health*, 483. For Goldschmidt's position toward National Socialism, see his autobiography: Richard Goldschmidt, *In and Out of the Ivory Tower* (Seattle: University of Washington Press, 1960).

20. Lothrop Stoddard, *The Rising Tide of Color Against White-Supremacy* (New

York: Charles Scribner's Sons, 1920); *The Revolt Against Civilization: The Menace of the Under-Man* (New York: Charles Scribner's Sons, 1922); *Racial Realities in Europe* (New York: Charles Scribner's Sons, 1924). See: Mehler, *History*, 428.

21. William L. Shirer, *Berlin Diary: The Journal of a Foreign Correspondent* (New York: Alfred Knopf, 1941): 257.
22. Stoddard, *Darkness*, 187.
23. Stoddard, *Darkness*, 189.
24. Stoddard, *Darkness*, 189.
25. Stoddard, *Darkness*, 190–91.
26. Stoddard, *Darkness*, 192–96; see also: Chase, *Legacy*, 347–51.

Chapter 6

1. Johnson, *International Eugenics*, 160.
2. Ludmerer, 148–50. Some recent scholarship also uses this distinction: Nils Roll-Hansen, "The Progress of Eugenics: Growth of Knowledge and Change in Ideology," *History of Science*, 26 (1988): 293–331.
3. It would be interesting to apply the concept of the social construction of knowledge to the controversies between supporters and critics of the Nazi race policy. Places to start would be: Ludwig Fleck, *Genesis and Development of a Scientific Fact*, trans. by F. Bradley and T. J. Trenn (Chicago: University of Chicago Press, [1935] 1979) and Karin Knorr-Cetina, *The Manufacture of Knowledge: An Essay on the Constructivist and Contextual Nature of Science* (New York: Pergamon Press, 1981). In *National Socialism and the Religion of Nature* (London & Sydney: Croom Helm, 1986), 72, Robert A. Pois has pointed out that by declaring scientific objectivity impossible, National Socialist ideology paralleled the criticism of the theory of the social construction of knowledge emerging from the work of Fleck in 1935. For interesting applications of the theory of the social construction of knowledge to the field of eugenics, see: Donald A. MacKenzie, *Statistics in Britain 1865–1930: The Social Construction of Scientific Knowledge* (Edinburgh: Edinburgh University Press, 1981) and Nicole Hahn Rafter, *White Trash: The Eugenic Family Studies, 1877–1919* (Boston: Northeastern, 1988).
4. This information is from Mehler, *History*, 134. The presidents of the AAAS were William H. Welch, David Starr Jordan, Charles W. Eliot, Henry Fairfield Osborn, and E. G. Conklin.
5. Babbott was a trustee of the Long Island Railroad, several New York elevated railways, and the Brooklyn Savings Bank. Garrett was a partner in Robert Garrett and Son, a banking firm, and the director of several other banks and insurance companies. Mehler, *History*, 135–36. Other sponsors of the American Eugenics Society were John D. Rockefeller and George Eastman.
6. Raymond Pearl, "The Biology of Superiority," *American Mercury*, 12 (1927): 260. See: Kevles, *Name*, 122–23.
7. Gregory to Pearl, May 6, 1935, and Pearl to Gregory, May 8, 1935, Pearl Papers. See: Allen, *Eugenics Record Office*, 253 and *Old Wine*, 255.
8. The complexity of Pearl's anti-Semitism needs further examination. For example, it is interesting that his friend H. L. Mencken published the attack by the anthropologist Franz Boas against the concept of an Aryan race.

9. Letter from Raymond Pearl to Franz Boas, October 3, 1935, Franz Boas Papers, American Philosophical Society, Philadelphia. See: Provine, *Geneticists*, 795 and Elazar Barkan, "Mobilizing Scientists against Nazi Racism, 1933–1939," *Bones, Bodies, Behavior: Essays on Biological Anthropology*, ed. George W. Stocking (Madison: University of Wisconsin Press, 1988): 185–86.

10. Raymond Pearl, "On the Incidence of Tuberculosis in the Offspring of Tuberculous Parents," *Zeitschrift für Rassenkunde und ihrer Nachbargebiete*, 3 (1937): 301–7.

11. The three existing studies about Davenport by Carletton MacDowell, Charles E. Rosenberg, and Daniel J. Kevles do not mention his relation to Nazi Germany. Carletton MacDowell, "Charles Benedict Davenport, 1866–1944: A Study of Conflicting Interests," *Bios*, 17 (1946): 3–50; Charles E. Rosenberg, "Charles Benedict Davenport and the Beginning of Human Genetic," *Bulletin of History of Medicine*, 35 (1961): 266–76, reprinted in Charles E. Rosenberg, *No Other Gods: On Science and American Social Thought* (Baltimore: Johns Hopkins University Press, 1976): 89–97; Kevles, *Name*, 41–56.

12. Charles B. Davenport, "Presidential Address: The Development of Eugenics," *A Decade of Progress*, 17–22. See also: Ludmerer, *Genetics*, 53.

13. Articles include: Charles B. Davenport, "The Influence of Economic Conditions on the Mixture of Races," *Zeitschrift für Rassenkunde*, 1 (1935): 17–19; Charles B. Davenport, "The Azygos Vein in Negros," *Zeitschrift für Rassenkunde*, 1 (1935): 82; Charles B. Davenport, "Eugenische Forschungen und ihre praktische Anwendung in den Vereinigten Staaten," *Der Erbarzt*, 29 (1936): 97–98; Charles B. Davenport, "Genetics of the Japanese," *Zeitschrift für Rassenkunde*, 3 (1937): 91. Publications in festschrifts: Charles B. Davenport, "How Early in Ontogeny Do Human Racial Characters Show Themselves?" *Eugen-Fischer-Festband, Zeitschrift für Morphologie und Anthropologie*, 34 (1934): 76–78 and Charles B. Davenport, "The Genetical Basis of Resemblance in the Form of the Nose," *Kultur und Rasse: Festschrift zum 60. Geburtstag Otto Reches* (Munich, Berlin: Lehmann, 1939): 60–64. Information about Reche in Weindling, *Health*, 500–7.

14. Just to Davenport, May 11, 1934, and cable from Davenport to Just, May 23, 1934, Davenport Papers: Just.

15. Julius Bauer to Davenport, June 24, 1934, Davenport Papers: Just.

16. Landauer to Davenport, October 19, 1935. The official reason Davenport gave for not signing the resolution was that he had resigned from the "Gesellschaft für Vererbungswissenschaft." Davenport to Landauer, October 23, 1935, Davenport Papers: Landauer.

17. Landauer to Davenport, February 29, 1936, and Davenport to Landauer, March 13, 1936, Davenport Papers: Landauer.

18. Bock, *Zwangssterilisation*, 63, defines eugenics as inherently racist. To trace the development of Bock's innovative concept of racism, see: Gisela Bock, "Antinatalism, Maternity and Paternity in National Socialist Racism," *Maternity and Gender Policies: Women and the Rise of the European Welfare State, 1880–1950s*, eds. Gisela Bock and Pat Thane (London: Routledge, 1991): 233–55. For the scholarship in the United States see: Garland Allen, "The Misuse of Biological Hierarchies: The American Eugenics Movement, 1900–1940," *History and Philosophy of the Life Sciences*, 5 (1983): 105–28; Robert Proctor, *Racial Hygiene*; Mehler, *History*; and Jerry Hirsch, "To 'Unfrock the Charlatans,'" *SAGE Race*

Relations Abstracts, 6/2 (1981): 1–65. See also a short note by Jerry Hirsch and Barry Mehler, "Eugenics has a long racist history," *Contemporary Psychology,* 31 (1986): 633.

19. Hans F. K. Günther, *Rassenkunde des deutschen Volkes* (Munich: Lehmann, ninth edition 1926): 14 (my translation).

20. Günther's understanding of race was based on the work of American anthropologist Madison Grant, whose book, *The Passing of the Great Race,* was a great success in the United States and appeared in a German translation in 1925.

21. Tilmann Broszat, *Zur Geschichte von Rassenhygiene/Eugenik und öffentlichen Gesundheitswesen vor und während der Zeit des Nationalsozialismus* (Munich: Gutachten im Auftrage des Instituts für Zeitgeschichte, 1983): 18–19.

22. Bock, *Zwangssterilisation,* 65.

23. Alfred Ploetz, "Die Begriffe Rasse und Gesellschaft und einige damit zusammenhängende Probleme," *Schriften der deutschen Gesellschaft für Soziologie,* 1 (1911), reprinted in *Archiv für Gesellschafts- und Rassenbiologie,* 28 (1934): 415–37; Alfred Ploetz, "Ziele und Aufgabe der Rassenhygiene," *Deutsche Vierteljahresschrift für öffentliche Gesundheitspflege,* 43 (1911), 164–92. His race ideology was first developed in: Alfred Ploetz, *Die Tüchtigkeit unserer Rasse und der Schutz der Schwachen* (Berlin, 1895). See also: Ignaz Kaup, "Ueber die eugenische Bewegung in England," *Concordia,* 18 (1911): 359–65.

24. Bock, *Zwangssterilisation,* 60. Similar argumentation can be found in: Michael Pollack, *Rassenwahn und Wissenschaft: Anthropologie, Biologie, Justiz und die nationalsozialistische Bevölkerungspolitik* (Frankfurt a.M.: Hain, 1990): 9–16.

25. Racial hygienists used "society" and "race" as synonyms. For example, see: Ploetz, *Begriff Rasse und Gesellschaft.* That concepts of race are social phenomena mapped into the social world, not the translation of "biology" into the social world, is stressed by several authors. See: Bock, *Zwangssterilisation,* 60; Nancy Stepan, *The Idea of Race in Science: Great Britain 1800–1960* (London: Macmillan, 1982): 47–82. For a more general discussion of racism see: Kurt Lenk, *"Volk und Staat": Strukturwandel politischer Ideologien im 19. und 20. Jahrhundert* (Stuttgart: Kohlhammer, 1971): 147–48; Heinz-Georg Marten, *Sozialbiologismus: Biologische Grundpositionen der politischen Ideengeschichte* (Frankfurt, New York: Campus, 1983): 147–48.

26. Popenoe, Johnson, *Applied Eugenics,* 1920, 298–314; Erwin Baur, Eugen Fischer, Fritz Lenz, *Grundriss der menschlichen Erblichkeitslehre und Rassenhygiene,* two vol. (Munich: Lehmann, first edition 1921, second edition 1923). See the definition of race in the fourth edition from 1936, vol. 1, 250.

27. Adolf Hitler, *Mein Kampf,* vol. 2 (Munich: Eher, 1928): 80–81. Quoted in: Bock, *Antinatalism,* 236.

28. The distinction has already been made implicitly in the studies of Haller and Ludmerer. For the application of this distinction to the Scandinavian eugenic movements, see: Nils Roll-Hansen, "Geneticists and the Eugenics Movement in Scandinavia," *The British Journal for the History of Science* 22 (1989): 335–46. For Great Britain, see Geoffrey R. Searle, *Eugenics and Politics in Britain, 1909–1914* (Leyden: Nordhoff International, 1976). Sheila F. Weiss applied a similar concept to the racial hygiene movement in Germany before 1933 in "The Race Hygiene Movement in Germany, 1904–1945, *The Wellborn Science: Eugenics in*

Germany, France, Brazil and Russia, ed. Mark B. Adams (New York, Oxford: Oxford University Press, 1990): 8–68.

29. Mehler, *History,* iii.
30. Kopp, *Eugenic Program,* 5.
31. Laughlin, *Eugenics in Germany,* 65–66.
32. Arthur Comte de Gobineau, *The Inequality of Human Races,* trans. from the French edition of 1853–1855 by A. Collins (London: Heinemann, 1915). For the development of Grant's and Stoddard's Nordic theory out of Gobineau's work, see: Dietrich Bronder, *Bevor Hitler kam* (Hannover: Hans Pfeiffer, 1964): 295. For a discussion concerning the close relationship of racial anthropologists to the organized eugenics movement in the United States, see: Hans-Peter Kröner, "Die Eugenik in Deutschland von 1891–1934," diss., University of Münster, 1980, 31. The more distant relation of German racial anthropologists to the racial hygiene movement is treated in: Hans Jürgen Lutzhöft, *Der Nordische Gedanke in Deutschland 1920 bis 1940* (Stuttgart: Klett-Cotta, 1971): 163–164 and Gunter Mann, "Rassenhygiene-Sozialdarwinismus," *Biologismus im 19. Jahrhundert,* ed. Günther Mann (Stuttgart: Ferdinand Enke, 1973): 73–77.
33. *The New York Times* (November 5, 1933): 16. In 1936 the official National Socialist recommendation for reading in the field of human heredity, racial hygiene, and population policy mentioned Grant's first book, *Der Untergang der grossen Rasse* (Munich: Lehmann, 1925), and Arthur Comte de Gobineau's *Essai sur l'inequalité des races humaines* (Paris, 1853–1855) as the only essential literature by non-German authors. *Rassenkunde: Eine Auswahl des wichtigsten Schrifttums aus dem Gebiet der Rassenkunde, Vererbungslehre, Rassenpflege und Bevölkerungspolitik,* ed. Institut für Lese- und Schrifttumskunde (Leipzig, 1936).
34. Memorandum to Miss Wycoff, October 26, 1933, by R. V. Coleman, Charles Scribner's Sons Papers, CO 101, series author file I, Box 67, Madison Grant folder, Princeton University Library. See: Chase, *Legacy,* 343.
35. Eugen Fischer, Vorwort, *Die Eroberung eines Kontinents oder die Verbreitung der Rassen in Amerika,* by Madison Grant, trans. Else Mez (Berlin: Alfred Metzner, 1937): vii. Quoted in Chase, *Legacy,* 343. Excerpts of Grant's *The Conquest of a Continent* were reprinted in a Berlin newspaper. "Hat Amerika es wirklich besser: Die Zukunft der nordischen Rasse, von Madison Grant," *Berliner Börsenzeitung* (February 20, 1938). An argument arose, however, between German and American racial anthropologists because Grant denied the racial superiority of Germans and belittled the influence of "German blood" in the American population. After the appearance of Grant's second book, on the "Nordic Settlement of America," National Socialists attacked him for underestimating the "fixed historical achievements of Germany." See: "Rassengeschichte Amerikas? Um das nordische Element in den USA," *Germania* (November 10, 1937). Grant's tendency to underemphasize Germany's role in race history was the reason why one German racial hygiene journal praised him and his Nordic ideas only with slight reservation at his death in 1937. See: note in *Zeitschrift für Rassenkunde,* 6 (1937): 272. *Volk und Rasse,* 12 (1937): 365, reported that the "Jewry had fought strongly against the courageous researcher" (my translation).
36. Grant to Laughlin, January 16, 1934, Laughlin Papers. Quoted after: Hassencahl, *Harry H. Laughlin,* 343–44.

37. Johnson, *International Eugenics*, 199–212, quote 212. Johnson was a leading figure of the American eugenics movement during the peak period of mainline eugenics influence, and he coauthored the primary eugenics textbook. But his argumentation closely resembled that of reform eugenicists. Johnson thus illustrates the difficulties entailed in categorizing eugenicists as either mainline or reformist. Curiously, the important role he played in outlining a theoretical distinction between these two groups of eugenicists has been entirely disregarded, even by historians who have focused on the differentiation between mainline and reform eugenicists. His name is not mentioned either in Ludmerer's *Genetics* or in Kevles' *Name*.
38. Mehler, *History*, 286–88.
39. Johnson, *International Eugenics*, 184–85 and 227.
40. Frederick Osborn, "Summary of the Proceedings of the Conference on Eugenics in Relation to Nursing," February 24, 1937, American Eugenics Society Papers: Conference on Eugenics in Relation to Nursing.
41. Annual Meeting of the American Eugenics Society, May 5, 1938, Osborn Papers: I, 10, APS, Philadelphia.

Chapter 7

1. Franz Boas in a letter to unknown, October 10, 1935, Boas Papers, APS: Boas.
2. Historians have seldom distinguished between socialist eugenicists and reform eugenicists. They have categorized all eugenicists who criticized the Nordic arrogance of mainline eugenicists and National Socialists as either only socialist or reform eugenicists. This approach, however, fails to recognize differences in the concept of race improvement held by the two groups. Their different reactions to Nazi race policies further show that the single distinction between reform eugenicists and mainline eugenicists does not cover the whole spectrum. See: Donald K. Pickens, *Eugenics and the Progressives* (Nashville: Vanderbilt University Press, 1968); Michael Freeden, "Eugenics and Progressive Thought: A Study in Ideological Affinity," *Historical Journal*, 22 (1979): 645–71; Diane Paul, "Eugenics and the Left," *Journal of the History of Ideas*, 45 (1984): 567–90. A comprehensive and important study of socialist eugenicists in Germany is Schwartz, *Sozialistische Eugenik*. See also: Michael Schwartz, "Sozialismus und Eugenik: Zur fälligen Revision eines Geschichtsbildes," *Internationale wissenschaftliche Korrespondenz zur Geschichte der deutschen Arbeiterbewegung*, 4 (1989): 465–89.
3. From a review of the English translation of the main eugenics and genetics textbook in Germany, Erwin Baur, Eugen Fischer, Fritz Lenz, *Human Heredity*, trans. Eden and Cedar Paul (New York: Macmillan, 1931): 21. Hermann J. Muller, "Human Heredity," *Birth Control Review*, 17 (1933): 21, reprinted in Hermann J. Muller, *Studies in Genetics* (Bloomington: Indiana University Press, 1962): 541–44.
4. Letter from Jennings to S. G. Levit, April 2, 1936, Jennings Papers: Levit.
5. Auswärtiges Amt, Ressortbesprechung at August 21, 1936, BAP 49.01, 2969.
6. Historians of eugenics differ widely in their interpretation of the *Manifesto*. Ludmerer, *Genetics*, 129, called it a "condemnation" of eugenics. Roll-Hansen, *Progress*, 312, saw in it a "formulation of the position of reform eugenics." Paul,

Eugenics, 583, claimed that the *Genetico Manifesto* is the "statement of socialist eugenic position." In my opinion, the content of the resolution, as well as the analysis of the signatories, can only confirm the last position.

7. The *Genetico Manifesto* was widely published and reprinted in the *Journal of Heredity*, 30 (1939): 371–73. The quote is from a reprint in Muller, *Studies*, 347. The Manifesto was originally signed by F. A. E. Crew, J. B. S. Haldane, S. C. Harland, L. T. Hogben, J. S. Huxley, H. J. Muller, and J. Needham. Later, the following added their names: G. P. Child, P. R. David, G. Dahlberg, T. Dobzhansky, R. A. Emerson, C. Gordon, J. Hammond, C. L. Huskins, W. Landauer, H. H. Plough, E. Price, J. Schultz, A. G. Steinberg, and C. H. Waddington. See the commentary of Weingart, Kroll, Bayertz, *Rasse*, 542, for arguments that the foreign critics of the German race policy were ambiguous because they remained committed to the eugenic credo and could therefore not be effective, either scientifically or morally.

8. Garland E. Allen, "The Eugenics Record Office at Cold Spring Harbor, 1910–1940: An Essay in Institutional History," *Osiris*, n.s., 2 (1986): 250; Kröner, *Eugenik*, 32; Ludmerer, *American Geneticists*, 350–54.

9. Quote from a letter from Boas to P. Baerwald, on June 12, 1933, Boas Papers: Baerwald. The following information is based on the article by Elazar Barkan, *Mobilizing*, 180–205. See also: Barkan, "Race Concepts in England and the United States between the two World Wars," diss., Brandeis University, 1988 and Carl N. Degler's *In the Search of Human Nature: The Decline and Revival of Darwinism in American Social Thought* (New York: Oxford University Press, 1991).

10. Letter from Boas to Baerwald, June 12, 1933, Boas Papers: Baerwald. See: Barkan, *Mobilizing*, 183.

11. Hooton in a letter to Grant on March 11, 1933, quoted according to Barkan, *Mobilizing*, 186. For Hooton's rejection of racial psychology see his book: *Twilight of Man* (New York: Putnam's Sons, 1939): 129. For his doubts that it is possible to clearly distinguish separate races, see: Hooton, *Up from the Ape* (New York: Macmillan, 1931).

12. Barkan, *Mobilizing*, 186

13. Barkan, *Mobilizing*, 186. He refers to Hooton's article, "Why the Jews Grow Stronger," *Collier's*, 103.5 (1939): 12–13 and 103.6 (1939): 71–72.

14. Letter from Hooton at March 2, 1938, quoted according to Barkan, *Mobilizing*, 186.

15. Hooton, "Ten Statements about Race," quoted according to Barkan, *Mobilizing*, 188.

16. Danforth to Hooton at April 11, 1935, quoted according to Barkan, *Mobilizing*, 189. Other anthropologists agreed but demanded major changes.

17. Boas in a letter to P. Rivet, on November 11, 1937, Boas Papers: Rivet. See: Barkan, *Mobilizing*, 197.

18. Osborn to Boas, October 11, 1937, Boas Papers: Frederick Osborn.

19. Resolution in a letter from Boas to Osborn, December 20, 1937. Osborn's agreement is in a letter to Boas dated December 23, 1937. About the relations between Boas and Hooton, see: Boas to unknown, October 18, 1937, Boas Papers.

20. More successful was Boas's initiative to organize American scientists to protest against an article of German Nobel Prize physicist Johannes Stark in which Stark,

as Boas reported to Hooton, distinguished physicists into "good, i.e., non theoretical and 'Aryan,' and bad, i.e., theoretical and Jewish." More than 1,000
scientists signed a resolution prepared by Franz Boas and a Columbia graduate
student named M. I. Finkelstein. Because American eugenicists played no role in
preparing or promoting the resolution, I do not go into greater details. For more
information, see: Barkan, *Mobilizing*, 197–200. For resolutions of the American
Association of University Professors, the American Anthropological Association,
and the Society for the Psychological Study of Social Issues, see the appendix of
the book of anthropologist Ruth Benedict, *Race: Science and Politics* (New York:
Viking Press, 1940).

21. "Report of the Advisory Committee of the Eugenics Record Office" (undated): 6,
    Laughlin Papers. New York geneticist L. C. Dunn was an important member of
    the Committee. Negative experiences during a visit in Nazi Germany in 1934 and
    1935 intensified his critical position toward eugenics. See Dunn's letter to John
    Merriam, July 3, 1935, Dunn Papers: Merriam. For more details about the Eu-
    genics Record Office, see the excellent article by Allen, *Eugenics*, 225–64.

22. Reilly, *Surgical Solution*, 96, reported 20,063 cases of sterilization from 1907 to
    1934 and 15,815 cases from 1935 to 1940. *Lancet* quoted statistics from the
    Journal of the American Medical Association stating that in the three-year period
    between 1941 and 1943, over 42,000 people were sterilized in the United States.
    "Sterilization of the Insane in the U.S.A.," *Lancet*, July 14, 1945: 63. See also:
    Phillip R. Reilly, "Involuntary Sterilization in the United States: A Surgical
    Solution," *The Quarterly Review of Biology*, 62 (1987): 153–70.

23. Quoted in Mehler, *History*, 118. Mehler has shown how Osborn focused his
    criticism of the old guard within the Society on Davenport. Osborn complained in
    1933 that Davenport overemphasized the relationship between eugenics and ge-
    netics and brought disrepute to the eugenics movement. Frederick Osborn,
    "Memorandum on the Eugenics Situation in the United States," May 24, 1933,
    American Eugenics Society Papers, 1.

24. Laughlin to Osborn, November 17, 1932, Laughlin Papers. Quoted in: Mehler,
    History, 115.

25. Osborn to Laughlin, approximately May 1937, Laughlin Papers. Quoted in:
    Mehler, *History*, 116.

26. Mehler, *History*, 289.

27. Frederick Osborn, "Social Implications of the Eugenic Program," *Child Study*,
    16 (1939): 95–97. See also: Frederick Osborn, "The American Concept of Eu-
    genics," *Eugenic News*, 24 (1939): 2.

28. Frederick Osborn, "Implications of the New Studies in Population and Psychol-
    ogy for the Development of Eugenic Philosophy," *Eugenic News*, 22 (1937): 106.
    See also: Frederick Osborn, "Memorandum on the Eugenics Situation in the
    United States," Osborn Papers: I # 2.

Chapter 8

1. Arthur Gütt, Ernst Rüdin, and Falk Ruttke, *Gesetz zur Verhütung erbkranken
   Nachwuchses vom 14. Juli 1933* (Munich: Lehmann, 1934): 13, quoted in:
   Weinreich, *Hitler's Professors*, 35–36.

2. Unpublished autobiography of Leon F. Whitney, written in 1971, Whitney Papers, APS, 204–5.
3. Chase, *Legacy*, 347. Geoffrey Hellmann, *Bankers, Bones & Beetles: The First Century of the American Museum of Natural History* (Garden City: The Natural History Press, 1969): 194.
4. Letter from Foster Kennedy to the Euthanasia Society, Archives of Choice in Dying, formerly Concern for Dying and the Society for the Right to Die, New York City. See also his letter to *The New York Times*, February 20, 1939. I thank Peter Lindley for guiding me to this source.
5. See Lifton, *Nazi Doctors*, 45. In a 1942 article, Kennedy expressed his support for the killing of "inferiors." Foster Kennedy, "The Problem of Social Control of the Congenital Defective: Education, Sterilization, Euthanasia," *American Journal of Psychiatry*, 99 (1942): 13–16. See also an earlier article: Foster Kennedy, "Sterilization and Eugenics," *American Journal of Obstetrics and Gynecology*, 34 (1937): 519–20. For an example of an article that revealed the mass killings in Nazi Germany to the American public, see: William L. Shirer, "Mercy Deaths in Germany," *Reader's Digest*, June 1941, 55–58.
6. Schneider to Laughlin, May 16, 1936 and Laughlin to Schneider, May 28, 1936, Laughlin Papers.
7. *The New York Times*, April 12, 1936; see Hassencahl, *Harry H. Laughlin*, 352.
8. See, for example, the letter of congratulations from Gosney of the Human Betterment Foundation, Laughlin Papers. For public acknowledgment, see: *Eugenic News*, 21 (1936): 79. "Rassenhygiene auch in den Vereinigten Staaten," *Völkischer Beobachter* (April 4, 1937); *Archiv für Rassen- und Gesellschaftsbiologie*, 30 (1936): 191.
9. Laughlin to Schneider, August 11, 1936, Laughlin Papers.
10. The quote from the diploma is in Hassencahl, *Laughlin*, 353–54.
11. An impressive example of the extensive use of positive statements from international congresses by Nazi propaganda is the International Congress for Criminal Law, which passed a resolution favoring compulsory eugenic sterilization. This resolution was used again and again for propaganda purposes by the National Socialists. See "Reichsärzteführer Dr. Gerhard Wagner zur deutschen Sterilisationsgesetzgebung: Sie ist international als mustergültig anerkannt," *Ziel und Weg*, 5 (1935): 440. The legitimation of castration in Germany was also partially based on a resolution of this congress and a resolution, passed by the International Federation of Eugenic Organizations in 1934. Arthur Gütt, "Bevölkerungspolitik als Aufgabe des Staates," *Bevölkerungsfragen: Bericht des Internationalen Kongresses für Bevölkerungswissenschaft Berlin 26.August–1.September 1935*, eds. Hans Harmsen and Franz Lohse (Munich: Lehmann, 1936): 749–50.
12. "Olympiagäste fragten, das Rassenpolitische Amt antwortete," *Neues Volk*, 4.9 (1936): 40. "Vortrag des Reichsamtsleiter Dr. Walter Gross bei dem Empfangsabend des Aussenpolitischen Amtes der NSDAP vor der ausländischen Diplomatie und Presse am 21. März 1935," *Rassenpolitische Auslandskorrespondenz*, 2.3 (1935), special edition.
13. Walter Gross, *Der deutsche Rassegedanke und die Welt* (Berlin: Junker and Dünnhaupt, 1939): 7 (my translation). A similar opinion is expressed in: Elizabeth Antonia Storch, "Mortalität und Morbidität bei eugenischen Sterilisierungen an

190 Frauen (ausgeführt im Städt. Krankenhaus in Speyer in der Zeit vom 24.4.35–1.1.39)," diss., University of Heidelberg, 1939, 5.

14. *Neues Volk* 2 (1934): 12–13 (my translation).

15. Institut für Zeitgeschichte, Munich, MA 1159 17404ff (my translation).

16. 2. Arbeitstagung des Rassenpolitischen Amtes der NSDAP, 2–9 June 1935, Institut für Zeitgeschichte, Munich, MA 1159 17448 (my translation).

17. "Vortrag des Reichsamtleiter Dr. Walter Gross bei dem Empfangsabend des Aussenpolitischen Amtes der NSDAP vor der ausländischen Diplomatie und Presse am 21. März 1935," *Rassenpolitische Auslandskorrespondenz*, 2.3 (1935), special edition.

18. "Welche Staaten bekämpfen Erbkrankheiten," *Völkischer Beobachter* (March 9, 1938).

19. " 'Rassismus' über Europa," *NSK* (August 26, 1937).

20. Schulz, *Das Gesetz*, 19–20. "Rassenhygiene auch in den Vereinigten Staaten," *Völkischer Beobachter* (April 4, 1937).

21. "Welche Staaten bekämpfen Erbkrankheiten," *Völkischer Beobachter* (March 9, 1938). See also: Wilhelm Jung, "Erwachendes Rassebewusstsein," *NS-Partei-Correspondence* (April 15, 1937).

22. Gross, *Rassegedanke*, 27, and E. Hester, "Deutschlands Rassenpolitik im Urteil der anderen: Amerikanische und englische Anerkennung für die deutsche Rassenpolitik." *Völkischer Beobachter* (August 19, 1936) (my translation).

23. The U.S. ambassador in Berlin reported to the secretary of state on December 16, 1938, that "the press persisted in its campaign to portray the United States to the German public as a nation in which all branches of activity including the Government are dominated and hence corrupted by the Jews." National Archives, Washington, D.C., 862.9111. See also: "Rassenhygiene auch in den Vereinigten Staaten," *Völkischer Beobachter* (April 4, 1937) and a satirical article in *Fredericus* (October 14, 1935).

24. See, for example, the poll of the Human Betterment Foundation which supposedly proved that well-educated and informed people were much more supportive of sterilizations in Germany. "Umfrage über die Sterilisation," *NSK* (July 28, 1936).

25. Interesting applications for the field race policy are Schmuhl's *Rassenhygiene* and Weindling's *Health*. The "policracy" discussion could benefit by giving more attention to the new approaches in political science, like policy analysis, regime theory, and multiple-elite theory.

26. For a more detailed description of these groups, see: Weindling, *Health*, 496.

27. Letter of the Reich Ministry of the Interior to Darré on May 23, 1934. The quoted statement of Fischer was included as copy in this letter. BAP RMI 15.01 26244 (my translation).

28. Quoted in the New York newspaper *PM*, August 21, 1945. See also Weinreich, *Hitler's Professors*, 33 and *Volk und Rasse*, 8 (1933): 158. Ernst Rüdin, "Aufgaben und Ziele der Deutschen Gesellschaft für Rassenhygiene," *Archiv für Rassen- und Gesellschaftsbiologie*, 28 (1934): 228–33.

29. Weindling, *Health*, 503, makes this point, but underestimates its importance for the standing of the organized racial hygienists within Nazi Germany.

30. Walter Gross, *Ziel und Weg*, 9 (1938): 535 (my translation). A similar statement

by Otmar Freiherr von Verschuer appears in "Rassenhygiene als Wissenschaft und Staatsaufgabe," *Der Erbarzt*, 3 (1936): 17–19.

31. Paul Popenoe, "The German Sterilization Law," *Journal of Heredity*, 25 (1934): 260.

Chapter 9

1. Karl Bonhoeffer, "Ein Rückblick auf die Auswirkungen und die Handhabung des nationalsozialistischen Sterilisationsgesetzes," *Der Nervenarzt*, 20 (1949): 2 (my translation).

2. The ambiguous position toward Nazi anti-Semitism is expressed for example in positions of Pearl, Hooton, Johnson, and Kopp.

3. Wilhelm Jung, "Wettkampf gegen den Rassetod: Die 'grausame deutsche Rassenlehre' und ihr Vergleich zum Ausland," *Preussische Zeitung* (March 21, 1937). One year before, *Volk und Rasse* printed a map that indicated all states in the United States that had passed laws against miscegenation between 1915 and 1930. *Volk und Rasse*, 8 (1936): 72. See also: "Schmelztiegel Amerika versagt," *Nationalsozialistische Landpost* (March 12, 1937).

4. Wilhelm Jung, "Erwachendes Rassebewusstsein," *Nationalsozialistische Parteikorrespondenz* (April 15, 1937).

5. "Doppelte Moral in USA," *Nationalsozialistische Parteikorrespondenz* (March 24, 1939).

6. "Rassentheorie und Rassenpraxis in USA," *Berliner Börsenzeitung* (February 22, 1939).

7. See, for example: *Rassenpolitische Auslandskorrespondenz*, 1.4 (1934). The measures in the United States, so argued the Nazi journal, against the "Nigger population . . . has been always stronger and more cruel than the German measures against the Jews will ever be able to be." See also: Walter Gross, "Die Welt und der Rassegedanke in Deutschland," *Zeitungsdienst* (November 2, 1934) and *Rassenpolitische Auslandskorrespondenz*, 8.5 (1935): 2–3.

8. "Das Rassenrecht in den Vereinigten Staaten," *Großdeutscher Pressedienst* (June 28, 1936) with reference on Heinrich Krieger, *Das Rassenrecht in den Vereinigten Staaten* (Berlin: Junker & Dünnhaupt, 1936).

9. Krieger, *Rassenrecht*, 10.

10. Krieger, *Rassenrecht*, 11. He quoted from Hans Grimm, "Amerikanische Briefe," *Münchener Neueste Nachrichten* (December 10, 1935) (my translation).

11. Discussion about a wartime program on November 11, 1942, Olden, *Human Betterment*, 189.

12. Bigelow was president of the American Eugenics Society from 1940 to 1946. Bigelow, *Brief History*, 49–51 and Osborn, *History*, 115–26.

13. Haller, *Eugenics*, 174, and Ludmerer, *Genetics*, 174.

14. Olden, *History*, 65. Whitney Papers: *Autobiography*, 204, APS.

15. Alexander Mitscherlich, Fritz Mielke, *The Death Doctors* (London: Elek, 1962), 256; Proctor, *Racial Hygiene*, 180.

16. Betty Booker, "Nazi Sterilizations had their roots in Eugenics," *Richmond Times-Dispatch* (February 27, 1980).

17. Mitscherlich, Mielke, *Death*, 248–49.

18. Karl Brand's defense exhibit no. 57, document no. 51, National Archives, Washington, D.C., Records of the United State Nuremberg War Crimes Trials, *United States* v. *Karl Brandt et al.* (Case 1), Record Group 238, microfilm roll #33, frames 0122–0124. Carrel's book is Karl Brand's defense exhibit no. 78, document no. 73, Archives Record Group 238, microfilm roll #42, Minute Book, Volume 31, 233. Ristow's book is Karl Brandt's defense exhibit, document no. 53. See: Lapon, *Mass-Murder,* 259.

19. Christian Pross, paper delivered to the conference "Medicine without Compassion" in Cologne, September 28–30, 1988, appeared in *The Nazi Doctors and the Nuremberg Code,* eds. M. Grodin and G. Annas (Oxford: Oxford University Press, 1992). See: Proctor, *Eugenics,* 177. See also: Bock, *Zwangssterilisation,* 104.

20. Schreiber, *Männer*; Paul Weindling, "Soziale Hygiene: Eugenik und medizinische Praxis-Der Fall Alfred Grotjahn," *Das Argument: Jahrbuch für kritische Medizin* (1984): 9; Weindling, *Health,* 569.

21. Mengele's academic work includes: Josef Mengele, "Sippenuntersuchungen bei Lippen-Kiefer-Gaumenspalten," *Zeitschrift für menschliche Vererbungs- und Konstitutionslehre,* 23 (1939): 2–42; "Die Vererbung der Ohrfisteln," *Der Erbarzt,* 8 (1940): 59–60. For more information about Mengele, see: "Die wissenschaftliche Normalität des Schlächters: Josef Mengele als Anthropologe 1937–1941," *Mitteilungen des Vereins zur Erforschung der nationalsozialistischen Gesundheits- und Sozialpolitik,* 1 (1985).

22. Verschuer Papers, University of Münster, quoted in: Müller-Hill, *Tödliche Wissenschaft,* 112.

23. For more information, see: Weingart, Kroll, Bayertz, *Geschichte,* 421–23. See also: Lucette Matalon Lagnado and Sheila Cohn Dekel, *Children of the Flames: Dr. Josef Mengele and the Untold Story of the Twins of Auschwitz* (London: Sidgwick & Jackson, 1991).

24. Miklos Nyiszli, *Auschwitz: A Doctor's Eyewitness Account* (New York: Frederick Fell, 1960). I thank Barry Mehler for guiding me to this source. See also: Lifton, *Nazi Doctors,* 305–51.

25. Lifton, *Nazi Doctors,* 358. See also: Müller-Hill, *Tödliche Wissenschaft,* 129. Müller-Hill based his evidence on an interview with Verschuer's son, Helmut Verschuer. Verschuer organized the relocation of the institute material from Berlin to Hessen, managing to see that his correspondence disappeared during the move. The contact from Mengele to Verschuer, however, is documented in letters that Mengele left in Auschwitz. See: Lilli Segal, *Die Hohenpriester der Vernichtung: Anthropologen, Mediziner und Psychiater als Wegbereiter von Selektion und Mord im Dritten Reich* (Berlin: Dietz, 1991): 173–75.

26. Von Verschuer to the *Neue Zeitschrift,* no date, Archive of the Max Planck Gesellschaft, Berlin, A2–II 56, quoted in: Weingart, Kroll, Bayertz, *Geschichte, 571.*

27. Verschuer to Muller, September 30, 1946 and July 31, 1947, Muller Papers, Indiana University, Bloomington. Quoted in: Weingart, Kroll, Bayertz, *Rassenhygiene,* 578–79.

28. Weingart, Kroll, Bayertz, *Rassenhygiene,* 579–80. One year later the German Society for Anthropology named Eugen Fischer an honorary member. See his

biography in Léon Poliakov and Josef Wulf, *Das Dritte Reich und seine Denker* (Berlin: Arani, 1959): 104.

29. Billig, *Rassistische Internationale*, 104. Pearson followed the British anthropologist Robert Gayre as editor of *The Mankind Quarterly* in 1978. Billig mentioned that two other scientists who played an important role in race science between 1933 and 1945 also contributed to *The Mankind Quarterly*. Before 1945, anthropologist Ilse Schwidetzky contributed regularly to the *Zeitschrift für Rassenkunde*; in 1961 she wrote an article for *The Mankind Quarterly* titled "Rassische Psychologie." In the 1930s Walter Scheidt wrote for the *Archiv für Rassen- und Gesellschaftsbiologie* and became, like Verschuer, a member of the editorial board of *The Mankind Quarterly*. For Scheidt's conflicted role in Nazi Germany, see: Horst Seidler, "Anthropologen im Widerstand?" *Der Widerstand gegen den Nationalsozialismus: Eine interdisziplinäre didaktische Konzeption zu seiner Erschliessung*, ed. Maria Zenner (Bochum: Brockmeyer, 1989): 67–122.

30. For more information: Paul Weindling, *Health*, 567–73, and Thomann, *Verschuer*, 63–65. Schade, who in 1936 summarized for the *Rassenpolitische Auslandskorrespondenz* the praise of non-German scientists for the Nazi race policy, became a professor in Düsseldorf. In 1972 he published "Gründe und Folgen des Geburtenrückganges," in the fascist *Nation Europa*, 22.11 (1972): 13–15. In 1976 he published in *Neue Anthropologie*, the German counterpart to *The Mankind Quarterly*, an article about the decline of the birthrate in Germany. Another member of the scientific advisory board of *Neue Anthropologie* was Arthur Jensen. See also: Heinrich Schade, *Völkerflut und Völkerschwund: Erkenntnisse und Mahnungen der Bevölkerungswissenschaft* (Berg am See: Kurt Vowinckel, 1973).

31. In *Nation Europa*, Günther used his nomination as a foreign member to the American Society for Human Genetics as an indication that he was always strictly scientific. *Nation Europa*, 11.6 (1961): 66.

32. See: Hans F. K. Günther (Ludwig Winter), *Begabtenschwund in Europa* (Pähl: Bebenburg 1959), and, *Vererbung und Umwelt* (fourth edition of his *Führeradel durch Sippenpflege: Vier Vorträge*, Munich: Lehmann 1936, second edition 1938, third edition 1941) (Pähl: Bebenburg, 1967). See also: *Mein Eindruck von Adolf Hitler* (Pähl: Bebenburg, 1969).

33. See: Billig, *Die Rassistische Internationale*, 64–65 and 95–101, as well as Segal, *Hohenpriester*, 187.

34. Bonhoeffer, *Ein Rückblick*, 1–5. Following the same logic, the American military tribunal in 1947 did not count the sterilization law among the Nazi crimes. Bock, *Zwangssterilisation*, 116. For Bonhoeffer's position on eugenics, sterilization, and euthanasia, see: Klaus-Jürgen Neumärker, *Karl Bonhoeffer: Leben und Werk eines deutschen Psychiaters und Neurologen in seiner Zeit* (Leipzig: Teubner, 1990): 140–55.

35. Harmsen, *Sterilization*, and Hans Nachtsheim, "Die Frage der Sterilisation vom Standpunkt der Erbbiologen," *Berliner Gesndheitsblatt*, 1 (1950): 603–4; *Für und Wider die Sterilisierung aus eugenischer Indikation* (Stuttgart: George Thieme, 1952), and "Das Gesetz zur Verhütung erbkranken Nachwuchses aus dem Jahre 1933 in heutiger Sicht," *Ärztliche Mitteilungen*, 33 (1962): 1640–44.

Chapter 10

1. Henry Sigerist, *Civilization and Disease* (Ithaca: Cornell University Press, 1943): 106–7. Quoted in Ludmerer, *Genetics*, 117.
2. See: Diane Paul, " 'Our Load of Mutations' Revisited," *Journal of the History of Biology*, 20 (1987): 321–35; Proctor, *Eugenics*, 192–93.
3. Olden, *From Birthright, Inc.*, 191.
4. Memo of Frederick Osborn from 1954, American Eugenics Society Papers: Pioneer Fund Foundation Grant. About the financing of Birthright, Inc. through the Pioneer Fund, see: Olden, *Birthright*, 274.

References

Manuscript Collections

Archiv des Diakonischen Werkes, Berlin (ADW) CA/GF
Archives of Choice in Dying, formerly Concern for Dying and the Society for the Right
    to Die, New York City
American Philosophical Society, Philadelphia (APS).
    American Eugenics Society Papers
    Franz Boas Papers
    Charles B. Davenport Papers
    L. C. Dunn Papers
    Herbert S. Jennings Papers
    Frederick Osborn Papers
    Raymond Pearl Papers
    Leon F. Whitney Papers
Bayerisches Hauptstaatsarchiv, Munich
    Bayrischer Minister des Innern
Bundesarchiv, Abteilung Potsdam (BAP)
    REM Reichserziehungsministerium
    RIM Reichsministerium des Innern
Bundesarchiv Koblenz (BAK)
    R 18 Reichsministerium des Innern
    R 43 Reichskanzlei
    NS 2 Rasse- and Siedlungshauptsamt-SS
Foundation Center, Washington
    Pioneer Fund
Indiana University Archive, Bloomington
    Hermann J. Muller Papers
Institut für Zeitgeschichte, Munich
Missouri State University, Kirksville
    Harry H. Laughlin Papers
National Archives, Washington
New York State Archive, Albany
    Film Collection
Princeton University
    Charles Scribner's Sons Papers
    Conklin Papers
Rockefeller Archive Center, North Tarrytown
University of Minnesota, Minneapolis
    Association for Voluntary Sterilization Papers

References

Journals, newspapers, and press agencies before 1945

Allgemeine Thüringische Landeszeitung
American Breeders Association, Annual Meetings
American Journal of Obstetrics and Gynecology
American Journal of Psychiatry
American Journal of Public Health
American Mercury
American Sociological Review
Angriff
Annals of the American Academy of Political and Social Science
Archiv für Kriminologie
Archiv für Rassen- und Gesellschaftsbiologie
Berliner Börsenzeitung
Berliner Tageblatt
Birth Control Review
Bulletin of the New York Academy of Medicine
Concordia
Deutsches Nachrichtenbüro
Deutsche Vierteljahresschrift für öffentliche Gesundheitspflege
Der Erbarzt
Eugenic News
Eugenics Review
Fortschritte der Erbpathologie, Rassenhygiene und ihrer Nachbargebiete
Fränkische Tageszeitung
Fredericus
Friedrich Mann's Pädagogisches Magazin
Germania
Grossdeutscher Pressedienst
Informationsdienst des Rassenpolitischen Amtes der NSDAP
Journal of Criminal Law, Criminology and Police Science
Journal of Heredity
Journal of Psycho-Asthenics
Medizin-Reform
Münchener Neueste Nachrichten
Nationalsozialistische Landpost
Nationalsozialistische Parteikorrespondenz
Neues Volk
The New York Times
Preussische Zeitung
Rassenpoltische Auslandskorrespondenz
Reader's Digest
Sächsische Staatszeitung
Schleswig-Holsteinische Hochschulblätter
Schriften der deutschen Gesellschaft für Soziologie
Time
Veröffentlichungen der Wiener Gesellschaft für Rassenpflege
Volk und Rasse

Völkischer Beobachter
Westfälische Zeitung
Wochenblatt Sachsen Anhalt
World's Work
Zeitschrift für die gesamte Neurologie und Psychiatrie
Zeitschrift für die gesamte Strafrechtswissenschaft
Zeitschrift für Morphologie und Anthropologie
Zeitschrift für Rassenkunde
Zeitungsdienst
Ziel und Weg

Publication before 1945

Authors who published before and after 1945 are mentioned in this section. Newspaper articles are not listed.

Bauer, Werner. "Erste Erfahrungen mit der Anwendung des Sterilisierungsgesetzes bei Geisteskranken," diss., University of Tübingen, 1936.

Baur, Erwin, Eugen Fischer, and Fritz Lenz. *Grundriss der menschlichen Erblichkeitslehre und Rassenhygiene*, two vol. (Munich: Lehmann, 1921, second edition 1923).

———, *Human Heredity*, trans. Eden and Cedar Paul (New York: Macmillan, 1931).

Benedict, Ruth. *Race: Science and Politics* (New York: Viking Press, 1940).

Berliner Gesellschaft für Rassenhygiene. *Die Leistung der Amerikaner auf rassenhygienischem Gebiete* (Berlin: Berliner Gesellschaft für Rassenhygiene, 1917 or 1918).

Besselmann, Paul Heinz. "Nationale eugenische Ausleseprobleme in der Volkswirtschaftslehre," diss., University of Heidelberg, 1934.

Blasbalg, Jenny. "Ausländische und deutsche Gesetzentwürfe über Unfruchtbarmachung," *Zeitschrift für die gesamte Strafrechtswissenschaft*, 52 (1932): 477–96.

Campbell, Clarence G. "The Biological Postulates of Population Study," *Bevölkerungsfragen: Bericht des Internationalen Kongresses für Bevölkerungswissenschaft Berlin 26.August–1.September 1935*, eds. Hans Harmsen and Franz Lohse (Munich: Lehmann, 1936): 601–11.

———, "Die deutsche Rassenpolitik," trans. Inge Bass, *Der Erbarzt* 3 (1936): 140–42.

———, "The German Racial Policy," *Eugenic News*, 21 (1936): 25–29.

———, "Urteil über die deutsche Rassenpolitik," *Rassenpolitische Auslandskorrespondenz* 3.10 (1936): 2–3.

Cook, Robert. "A Year of German Sterilization," *Journal of Heredity*, 26 (1935): 485–89.

Czech-Jochberg, Erich. *Adolf Hitler und sein Stab* (Oldenburg: Gerd Stalling, 1933).

Davenport, Charles B. "How Early in Ontogeny Do Human Racial Characters Show Themselves?," *Eugen-Fischer-Festband, Zeitschrift für Morphologie und Anthropologie*, 34 (1934): 76–78.

———, "The Azygos Vein in Negroes," *Zeitschrift für Rassenkunde*, 1 (1935): 82.

———, "The Influence of Economic Conditions on the Mixture of Races," *Zeitschrift für Rassenkunde*, 1 (1935): 17–19.

References

———, "Eugenische Forschungen und ihre praktische Anwendung in den Vereinigten Staaten," *Der Erbarzt*, 29 (1936): 97–98.

———, "Genetics of the Japanese," *Zeitschrift für Rassenkunde*, 3 (1937): 91.

———, "The Genetical Basis of Resemblance in the Form of the Nose," *Kultur und Rasse: Festschrift zum 60. Geburtstag Otto Reches* (Munich, Berlin: Lehmann, 1939): 60–64.

Ellinger, T.U.H. "On the Breeding of Aryans and Other Genetic Problems of War-Time Germany," *Journal of Heredity*, 33 (1942): 141–43.

Eugenics Education Society. *The First International Eugenic Congress, London* (London: Eugenics Education Society, n.d.).

Feilchenfeld. "Die Bestrebung der Eugenik in den Vereinigten Staaten von Nordamerika und ihre Übertragung auf deutsche Verhältnisse," *Medizin-Reform*, 21 (1913): 477–82.

Fischer, Eugen. "Schlussansprache," *Bevölkerungsfragen: Bericht des Internationalen Kongresses für Bevölkerungswissenschaft Berlin 26.August–1.September 1935*, eds. Hans Harmsen and Franz Lohse (Munich: Lehmann, 1936): 927–31.

Fleck, Ludwig, *Genesis and Development of a Scientific Fact*, trans. F. Bradley and T. J. Trenn (Chicago: University of Chicago Press, [1935] 1979).

Fortpflanzung, Vererbung, Rassenhygiene: Katalog der Gruppe Rassenhygiene der Internationalen Hygiene Ausstellung 1911 in Dresden, eds. Max von Gruber and Ernst Rüdin (Munich: Lehmann, 1911).

Frercks, Rudolf. *Deutsche Rassenpolitik* (Leipzig: Reclam, 1937).

Galton, Francis. *Inquiries into Human Faculty* (London: Macmillan, 1883).

Gaupp, Robert. *Die Unfruchtbarmachung geistig und sittlich Kranker und Minderwertiger* (Berlin: Julius Springer, 1925).

Gobineau, Arthur Comte de. *Essai sur l'inequalité des races humaines* (Paris, 1853–1855).

Goddard, Henry Herbert. *The Kallikak Family: A Study in the Heredity of Feeble-Mindedness* (New York: Macmillan, 1912).

———, *Feeble-Mindedness: Its Causes and Consequences* (New York: Macmillan, 1914).

———, "Feeblemindedness," *Journal of Psycho-Asthenics*, 33 (1928): 219–27.

Goethe, Charles M. "Patriotism and Racial Standards," *Eugenic News*, 21 (1936): 65–69.

———, "Eugenics and Geography," *Eugenic News*, 22 (1937): 47.

———, "Extinction of the Inca Highcastes: Presidential Address before the Twenty-Fifth Annual Meeting of the Eugenics Research Association," *Eugenic News*, 22 (1937): 51–57.

Goldschmidt, Richard. "Anthropological Determination of 'Aryanism,'" *Journal of Heredity*, 33 (1942): 215–16.

———, *In and Out of the Ivory Tower* (Seattle: University of Washington Press, 1960).

Gosney, Eugene S., and Paul Popenoe. *Sterilization for Human Betterment* (New York: Macmillan, 1929).

———, *Sterilisierung zum Zwecke der Aufbesserung des Menschengeschlechts*, trans. Konrad Burchardi (Berlin: Marcus & Webers, 1930).

Grant, Madison. *Der Untergang der großen Rasse,* trans. Else Mez (Munich: Lehmann, 1925).

———, *Die Eroberung eines Kontinents oder die Verbreitung der Rassen in Amerika,* trans. Else Mez (Berlin: Alfred Metzner, 1937).

Gross, Walter. "Die Welt und der Rassengedanke in Deutschland," *Zeitungsdienst,* November 2, 1934).

———, *Der deutsche Rassegedanke und die Welt* (Berlin: Junker and Dünnhaupt, 1939).

Günther, Franz. "Das neue Reichsgesetz gegen gefährliche Gewohnheitsverbrecher," *Archiv für Kriminologie,* 93 (1933): 197–200.

Günther, Hans F. K. *Rassenkunde des deutschen Volkes* (Munich: Lehmann, ninth edition 1926).

———, *Führeradel durch Sippenpflege: Vier Vorträge* (Munich: Lehmann, 1936, second edition 1938, third edition 1941).

———, (Ludwig Winter), *Begabtenschwund in Europa* (Pähl: Bebenburg, 1959).

———, *Vererbung und Umwelt* (Pähl: Bebenburg, 1967).

———, *Mein Eindruck von Adolf Hitler* (Pähl: Bebenburg, 1969).

Gütt, Arthur. "Bevölkerungspolitik als Aufgabe des Staates," *Bevölkerungsfragen: Bericht des Internationalen Kongresses für Bevölkerungswissenschaft Berlin 26.August–1.September 1935,* eds. Hans Harmsen and Franz Lohse (Munich: Lehmann, 1936): 745–56.

———, Ernst Rüdin, and Falk Ruttke. *Gesetz zur Verhütung erbkranken Nachwuchses vom 14. Juli 1933* (Munich: Lehmann, 1934).

Hager, Frithjof. "Der gegenwärtige Stand der Frage der Sterilisierung Minderwertiger in Deutschland," diss., University of Kiel, 1934.

Hartmann, Hans. "Die deutsche erbbiologische Forschung: Zum 25 jährigen Bestehen der Kaiser-Wilhelm-Gesellschaft zur Förderung der Wissenschaft," *Der Erbarzt,* 3 (1936): 3–8.

Hitler, Adolf. *Mein Kampf* (Munich: Eher, 1928).

———, *Mein Kampf,* intro. D. C. Watt (London, 1974).

Hoffmann, Géza von. *Rassenhygiene in den Vereinigten Staaten von Nordamerika* (Munich: Lehmann, 1913).

———, "Rassenhygienische Jahresversammlung in den Vereinigten Staaten von Nordamerika," *Archiv für Rassen- und Gesellschaftsbiologie,* 10 (1913): 829–30.

———, "Die rassenhygienischen Gesetze des Jahres 1913 in den Vereinigten Staaten," *Archiv für Rassen- und Gesellschaftsbiologie,* 11 (1914): 21–32.

———, "Das Sterilisierungsprogramm in den Vereinigten Staaten von Nordamerika," *Archiv für Rassen- und Gesellschaftsbiologie,* 11 (1914): 184–92.

———, *Krieg und Rassenhygiene: Die Bevölkerungspolitische Aufgabe nach dem Kriege* (Munich: Lehmann, 1916).

Holzmann, Erika. "Erfahrungen und Ergebnisse der Untersuchungen auf Ehetauglichkeit in Hamburg vom 20. Oktober 1935 bis 1. Juli 1940," diss., University of Rostock, 1941.

Hooton, Earnest A. *Up from the Ape* (New York: Macmillan, 1931).

———, *Twilight of Man* (New York: Putnam's Sons, 1939).

———, "Why the Jews Grow Stronger," *Collier's,* 103.5 (1939): 12–13 and 103.6 (1939): 71–72.

Hüllstrung, Herbert. "Über gesetzliche Bestimmungen und Erfolge der Zwangs-sterilisierung und Zwangskastration," diss., University of Bonn, 1934.

Human Betterment Foundation, *Human Sterilization* (Pasadena, Cal.: Human Betterment Foundation, n.d.).

―――, *Human Sterilization Today* (Pasadena, Cal.: Human Betterment Foundation, 1939).

Huntington, Ellsworth. *Tomorrow's Children: The Goal of Eugenics* (New York: John Wiley, 1935).

International Federation of Eugenic Organizations. *Bericht über die 11. Versammlung der Internationalen Föderation eugenischer Organisationen, Konferenz-sitzungen vom 18. bis 21. Juli 1934 im Waldhaus Dolden, Zurich* (Zurich: O. Füssli, 1934).

―――, *Bericht der 12. Versammlung der Internationalen Federation Eugenischer Organisationen, Konferenzsitztungen vom 15. bis 20. Juli 1936, Scheveningen, Holland* (Den Haag: W. P. van Stockum, 1936).

Jennings, Herbert S. *The Biological Basis of Human Nature* (New York: W. W. Norton, 1930).

Johnson, Roswell H. "International Eugenics," diss., University of Pittsburgh, 1934.

Kankeleit, Otto. "Künstliche Unfruchtbarmachung aus rassenhygienischen und sozialen Gründen," *Zeitschrift für die gesamte Neurologie und Psychiatrie*, 98 (1925): 220–53.

―――, *Die Unfruchtbarmachung aus rassenhygienischen und sozialen Gründen* (Munich: Lehmann, 1929).

―――, "Die Ausschaltung geistig Minderwertiger von der Fortpflanzung," *Volk und Rasse*, 6 (1931): 174–79.

Kaup, Ignaz. "Ueber die eugenische Bewegung in England," *Concordia*, 18 (1911): 359–65.

Kennedy, Foster. "Sterilization and Eugenics," *American Journal of Obstetrics and Gynecology*, 34 (1937): 519–20.

―――, "The Problem of Social Control of the Congenital Defective: Education, Sterilization, Euthanasia," *American Journal of Psychiatry*, 99 (1942): 13–16.

Knapp, W. "Statistisches und Empirisches über das Gesetz zur Verhütung erbkranken Nachwuchses," diss., University of Bonn, 1934.

Kopp, Marie E. "Legal and Medical Aspects of Eugenic Sterilization in Germany," *American Sociological Review*, 1 (1936): 761–70.

―――, "The German Program of Marriage Promotion through State Loan," *Eugenic News*, 21 (1936): 121–29.

―――, "Eugenic Sterilization Laws in Europe," *American Journal of Obstetrics and Gynecology*, 34 (1937): 499–505.

―――, "Eugenic Sterilization Laws in Europe," *Bulletin of the New York Academy of Medicine*, 34 (1937): 761–69.

―――, "Surgical Treatment of Sex Crime: Prevention Measure," *Journal of Criminal Law, Criminology and Police Science*, 28 (1938): 692–706.

Krause, Ursula. "Erfahrungen und Ergebnisse bei 315 Sterilisationen aus eugenischer Indikation·vom 4. April 1934 bis zum 1. April 1936," diss., University of Kiel, 1937.

Krauss, Hans. *Die Grundgedanken der Erbkunde und Rassenhygiene in Frage und Antwort* (Munich: Otto Gmelin, 1935).

Kreienberg, Walter. "Die Auswirkungen des Gesetzes zur Verhütung erbkranken Nachwuchses an dem Krankenbestand der Psychiatrischen und Nervenklinik Erlangen," diss., University of Erlangen, 1937.

Krieger, Heinrich. *Das Rassenrecht in den Vereinigten Staaten* (Berlin: Junker & Dünnhaupt, 1936).

Kuhlberg, Helmut. "Die Auswirkung des Gesetzes zur Verhütung erbkranken Nachwuchses in der Heil- und Pflegeanstalt Waldbröl," diss., University of Bonn, 1934.

Kürten, Heinz. "Der Kongress der Internationalen Föderation Eugenischer Organisationen," *Ziel und Weg,* 4 (1934): 15–16.

Landman, Jacob H. *Human Sterilization: The History of the Sexual Sterilization Movement* (New York: Macmillan, 1932).

Laughlin, Harry H. *The Legal, Legislative and Administrative Aspects of Sterilization: Report of the Committee to Study and to Report on the Best Practical Means of Cutting Off the Defective Germ-plasm in the American Population* (Cold Spring Harbor: Eugenics Record Office, 1914).

———, *Eugenic Sterilization in the United States* (Chicago: Psychopathic Laboratory of the Municipal Court of Chicago, 1922).

———, "Die Entwicklung der gesetzlichen rassenhygienischen Sterilisierung in den Vereinigten Staaten," *Archiv für Rassen- und Gesellschaftsbiologie,* 21 (1929): 253–62.

———, "The Legal Status of Sterilization," *Supplement to the Annual Report of the Municipal Court of Chicago for the Year 1929.*

———, "Studies on the Historical and Legal Development of Eugenic Sterilization in the United States," *Bevölkerungsfragen: Bericht des Internationalen Kongresses für Bevölkerungswissenschaft Berlin 26.August–1.September 1935,* eds. Hans Harmsen and Franz Lohse (Munich: Lehmann, 1936): 664–75.

———, "Eugenics in Germany: Motion Picture Showing How Germany Is Presenting and Attacking Her Problems in Applied Eugenics," *Eugenic News,* 22 (1937): 65–66.

Lenz, Fritz. "Eugenics in Germany," trans. Paul Popenoe, *Journal of Heredity,* 15 (1924): 223–31.

———, *Menschliche Auslese und Rassenhygiene (Eugenik)* (Munich: Lehmann, third edition 1931).

Mengele, Josef. "Sippenuntersuchungen bei Lippen-Kiefer-Gaumenspalten," *Zeitschrift für menschliche Vererbungs- und Konstitutionslehre,* 23 (1939): 2–42.

———, "Die Vererbung der Ohrfisteln," *Der Erbarzt,* 8 (1940): 59–60.

Muller, Hermann J. "Human Heredity," *Birth Control Review,* 17 (1933): 19–21.

———, *Studies in Genetics* (Bloomington: Indiana University Press, 1962).

Neeff, Dora. "Die bisherigen Erfahrungen über Eingriff und Verlauf der sterilisierenden Operation bei der Frau," diss., University of Heidelberg, 1935.

Norton, Marion S. [Marian S. Olden]. *Selective Sterilization in Primer Form* (Princeton: Sterilization League of New Jersey, 1937).

———, *Major Provisions for Population Control Abroad* (Princeton: Sterilization League of New Jersey, 1939).

———, *Human Betterment Was Our Goal* (Gwynedd, 1970).

———, *From Birthright, Inc., to Voluntary Sterilization* (Gwynedd, 1970).

———, *Sterilization League of New Jersey, 1937–1942* (Gwynedd, 1970).

References

Osborn, Frederick. "Implications of the New Studies in Population and Psychology for the Development of Eugenic Philosophy," *Eugenic News*, 22 (1937): 104–6.

————, "The American Concept of Eugenics," *Eugenic News*, 24 (1939): 2.

————, "Social Implications of the Eugenic Program," *Child Study*, 16 (1939): 95–97.

————, "History of the American Eugenics Society," *Social Biology*, 21 (1974): 115–26.

Osterfeld, Theo. "Über die Sterilisation aus eugenischer Indikation," diss., University of Würzburg, 1936.

Paulsen, Jens. "Politische Entwicklungsrichtungen in den europäischen Völkern," *Archiv für Rassen und Gesellschaftsbiologie*, 33 (1939): 224–31.

Pearl, Raymond. "Breeding Better Men," *World's Work*, 15 (1908): 9819–24.

————, "The Biology of Superiority," *American Mercury*, 12 (1927): 257–60.

————, "On the Incidence of Tuberculosis in the Offspring of Tuberculous Parents," *Zeitschrift für Rassenkunde und ihrer Grenzgebiete*, 3 (1937): 301–7.

Peter, William W. "Ein amerikanischer Arzt zur Sterilisierung in Deutschland," *Neues Volk*, 2 (1934): 16–18.

————, "German Sterilization Program," *American Journal of Public Health*, 24 (1934): 187–91.

Ploetz, Alfred. *Die Tüchtigkeit unserer Rasse und der Schutz der Schwachen* (Berlin, 1895).

————, "Die Begriffe Rasse und Gesellschaft und einige damit zusammenhängende Probleme," *Schriften der deutschen Gesellschaft für Soziologie*, 1 (1911), reprinted in *Archiv für Gesellschafts- und Rassenbiologie*, 28 (1934): 415–37.

————, "Ziele und Aufgabe der Rassenhygiene," *Deutsche Vierteljahresschrift für öffentliche Gesundheitspflege*, 43 (1911), 164–92.

Popenoe, Paul. "Rassenhygiene (Eugenik) in den Vereinigten Staaten," trans. Fritz Lenz, *Archiv für Rassen- und Gesellschaftsbiologie*, 15 (1923/1924): 184–93.

————, "Rassenhygienische Sterilisierung in Kalifornien," *Archiv für Rassen- und Gesellschaftsbiologie*, 23 (1931): 249–59.

————, "The German Sterilization Law," *Journal of Heredity*, 25 (1934): 257–60.

————, "The Progress of Eugenic Sterilization," *Journal of Heredity*, 25 (1934): 19–25.

————, "Trends in Human Sterilization," *Eugenic News*, 22 (1937): 42–43.

————, and Eugene S. Gosney, *Twenty-eight Years of Sterilization in California* (Pasadena, Cal.: Human Betterment Foundation, 1939).

————, and Roswell H. Johnson. *Applied Eugenics* (New York: Macmillan, 1920, second edition 1933, third edition 1939).

Problems in Eugenics: Papers Communicated to the First International Eugenics Congress held at the University of London, July 24th to 30th, 1912 (London: Eugenics Education Society, 1912).

Rassenkunde: Eine Auswahl des wichtigsten Schrifttums aus dem Gebiet der Rassenkunde, Vererbungslehre, Rassenpflege und Bevölkerungspolitik, ed. Institut für Lese- und Schrifttumskunde (Leipzig, 1936).

Reche, Otto. "Die Bedeutung der Rassenpflege für die Zukunft unseres Volkes," *Veröffentlichungen der Wiener Gesellschaft für Rassenpflege*, 1 (1925): 6.

Ross, Edward A. "The Causes of Race Superiority," *Annals of the American Academy of Political and Social Science*, 18 (1901): 85–88.

Rüdin, Ernst. "Aufgaben und Ziele der Deutschen Gesellschaft für Rassenhygiene," *Archiv für Rassen- und Gesellschaftsbiologie*, 28 (1934): 228–33.

————, "Strömungen in Art und Umfang der Sterilisationspraxis," *Archiv für Rassen- und Gesellschaftsbiologie*, 31 (1937): 367–69.

Ruttke, Falk. "Erbpflege in der Deutschen Gesetzgebung," *Ziel und Weg*, 4 (1934): 600–3.

————, "Erbpflege in der deutschen Gesetzgebung," *Der Erbarzt*, 3 (1936): 111–17.

Schade, Heinrich. "Der International Kongreß für Bevölkerungswissenschaft in Berlin," *Der Erbarzt* 2 (1935): 140–41.

————, "Ausländische Stimmen zur deutschen Erb- und Rassenpflege," *Rassenpolitische Auslandskorrespondenz*, 3.5 (1936): 3–4.

————, "Gründe und Folgen des Geburtenrückganges," *Nation Europa*, 22.11 (1972): 13–15.

————, *Völkerflut und Völkerschwund: Erkenntnisse und Mahnungen der Bevölkerungswissenschaft* (Berg am See: Kurt Vowinckel, 1973).

Schneidewind, H. H. von. *Wirtschaft und Wirtschaftspolitik der Vereinigten Staaten von Amerika* (Würzburg: Konrad Triltsch, 1933).

Schopohl, Heinrich. "Die Eugenik im Dienste der Volkswohlfahrt," *Volkswohlfahrt*, 13 (1932): 469–587.

Schulz, Reimer. "Das Gesetz zur Verhütung erbkranken Nachwuchses im Spiegel amerikanischer und deutschamerikanischer Presse," *Schleswig-Holsteinische Hochschulblätter*, 34 (1934): 19–20.

Shirer, William L. *Berlin Diary: The Journal of a Foreign Correspondent* (New York: Alfred A. Knopf, 1941).

————, "Mercy Deaths in Germany," *Reader's Digest*, June 1941, 55–58.

Sigerist, Henry. *Civilization and Disease* (Ithaca: Cornell University Press, 1943).

Steinwallner, Bruno. "Rassenhygienische Gesetzgebung und Maßnahmen ausmerzender Art," *Fortschritte der Erbpathologie, Rassenhygiene und ihrer Grenzgebiete*, 1 (1937): 193–260.

Stoddard, Lothrop. *The Rising Tide of Color Against White-Supremacy* (New York: Charles Scribner's Sons, 1920).

————, *The Revolt Against Civilization: The Menace of the Under-Man* (New York: Charles Scribner's Sons, 1922).

————, *Racial Realities in Europe* (New York: Charles Scribner's Sons, 1924).

————, *Into the Darkness: Nazi Germany Today* (New York: Deull, Sloan & Pearce, 1940).

Storch, Elizabeth Antonia. "Mortalität und Morbidität bei eugenischen Sterilisierungen an 190 Frauen (ausgeführt im Städt. Krankenhaus in Speyer in der Zeit vom 24.4.35–1.1.39)," diss., University of Heidelberg, 1939.

Striehn, Otto. "Kastration," diss., University of Munich, 1938.

Tietze, Felix. "Sterilisierung zum Zwecke der Aufbesserung des Menschengeschlechts," *Archiv für Rassen- und Gesellschaftsbiologie*, 25 (1931): 346–47.

Thomalia, C. "The Sterilization Law in Germany," trans. Alice Hellmer, *Eugenic News*, 19 (1934): 137–40.

Verschuer, Otmar Freiherr von. "Rassenhygiene als Wissenschaft und Staatsaufgabe," *Der Erbarzt*, 3 (1936): 17–19.

Wappler, Paul. "Über die gesetzliche Sterilisation und unsere Erfahrungen hiermit an Hand von 220 Sterilisierungen," diss., University of Leipzig, 1937.

References

Whitney, Leon F. *The Case for Sterilization* (New York: Frederick A. Stocks, 1934).

Woodruff, Charles. "Climate and Eugenics," *American Breeders Association, Proceedings of Annual Meetings,* 6 (1910): 122.

Publication after 1945

Adams, Mark B. "Eugenics in Russia 1900–1940," *The Wellborn Science: Eugenics in Germany, France, Brazil, and Russia,* ed. Mark B. Adams (New York, Oxford: Oxford University Press, 1990): 153–216.

Allen, Garland E. "The Misuse of Biological Hierarchies: The American Eugenics Movement, 1900–1940," *History and Philosophy of the Life Sciences,* 5 (1983): 105–28.

———, "The Eugenics Record Office at Cold Spring Harbor, 1910–1940: An Essay in Institutional History," *Osiris,* n.s., 2 (1986): 225–64.

———, "Old Wine in New Bottles: From Eugenics to Population Control in the Work of Raymond Pearl," *The Expansion of American Biology,* eds. Keith R. Benson, Jane Maienschein, and Ronald Rainger (New Brunswick, London: Rutgers University Press, 1991): 231–61.

———, and Barry Mehler. "Sources in the Study of Eugenics #1: Inventory of the American Eugenics Society Papers," *Mendel Newsletter, Archival Resources for the History of Genetics and Allied Sciences,* 14 (1977): 9–15.

Altner, Günther. *Weltanschauliche Hintergründe der Rassenlehre des Dritten Reiches: Zum Problem einer umfassenden Anthropologie* (Zurich: EVZ, 1968).

Alvarez-Peláez, R. "El Instituto de Medicina Social: Primeros Intentos de Institucionalizar la Eugenesia (en Espana)," *Acta Ill Congreso Sociedad Espanola de Historia de las Ciencias,* Octubre 1984.

———, "Introduction al estudio de la Eugenesia espanola (1900–1936)," *Quipu* 2 (1985): 95–122.

Arvey, Richard D., Thomas J. Bouchard, Nancy L. Segal, and Lauren M. Abraham, "Job Satisfaction: Environmental and Genetic Components," *Journal of Applied Psychology,* 74 (1989): 187–92.

Bajema, Carl L., ed., *Eugenics: Then and Now* (Stroudsburg: Hutchinson & Ross, 1976).

Baker-Benfield, G. J. *The Horrors of the Half-Known Life: Male Attitudes Toward Women and Sexuality in Nineteenth-Century America* (New York: Harper Colophon, 1976).

Barkan, Elazar. "Mobilizing Scientists against Nazi Racism, 1933–1939," *Bones, Bodies, Behavior: Essays on Biological Anthropology,* ed. George W. Stocking (Madison: University of Wisconsin Press, 1988): 180–205.

———, "Race Concepts in England and the United States between the Two World Wars," diss., Brandeis University, 1988.

Bergmann, Anna, Gabriele Czarnowski and Annegret Ehmann. "Menschen als Objekte humangenetischer Forschung und Politik im 20. Jahrhundert: Zur Geschichte des Kaiser-Wilhelm-Instituts für Anthropologie, menschliche Erblehre und Eugenik in Berlin–Dahlem (1927–1945)," *Der Wert des Menschen: Medizin in Deutschland 1918–1945,* ed. Ärztekammer Berlin (Berlin: Edition Hentrich, 1989): 121–42.

Bigelow, Maurice A. "Brief History of the American Eugenics Society," *Eugenic News*, 31 (1946): 49–51.

Billig, Michael. *Psychology, Racism and Fascism* (Birmingham: Searchlight, 1979).

———, "Professor Eysenck's Political Psychology," *Patterns of Prejudice*, 13.5 (1979): 9–16.

———, *Die Rassistische Internationale* (Frankfurt a.M.: Neue Kritik, 1981).

Bock, Gisela. *Zwangssterilisation im Nationalsozialismus: Studien zur Rassenpolitik und Frauenpolitik* (Opladen: WDV, 1986).

———, "Antinatalism, Maternity and Paternity in National Socialist Racism," *Maternity and Gender Policies: Women and the Rise of the European Welfare State, 1880s–1950s*, eds. Gisela Bock and Pat Thane (London: Routledge, 1991): 233–55.

Bonhoeffer, Karl. "Ein Rückblick auf die Auswirkungen und die Handhabung des nationalsozialistischen Sterilisationsgesetzes," *Der Nervenarzt*, 20 (1949): 1–5.

Bouchard, Thomas J., Auke Tellegen, David T. Lykken, and Kimerly J. Wilcox. "Personality Similarity in Twins Reared Apart and Together," *Journal of Personality and Social Psychology*, 54 (1988): 1031–39.

———, David T. Lykken, Matthew McGue, Nancy L. Segal, and Auke Tellegen, "Sources of Human Psychological Differences: The Minnesota Study of Twins Reared Apart," *Science*, 250.4978 (1990): 223–28.

———, and Matthew McGue. "Genetic and Rearing Environmental Influences on Adult Personality: An Analysis of Adopted Twins Reared Apart," *Journal of Personality*, 58 (1990): 263–93.

Brody, Nathan, Michael C. Corballis, Linda S. Gottfredson, William Shockley, and Arthur R. Jensen. "Commentary on Arthur R. Jensen (1955), The Nature of Black–White Differences on Various Psychometric Tests: Spearman's Hypothesis," *Behavioral and Brain Science*, 10 (1987): 507–18.

Bronder, Dietrich. *Bevor Hitler kam* (Hannover: Hans Pfeiffer, 1964).

Broszat, Tilmann. *Zur Geschichte von Rassenhygiene/Eugenik und öffentlichen Gesundheitswesen vor und während der Zeit des Nationalsozialismus* (Munich: Gutachten im Auftrage des Instituts für Zeitgeschichte, 1983).

Burleigh, Michael. "'Euthanasia' and the Cinema in Nazi Germany," *History Today*, 40 (1990): 11–16.

Chase, Allen. *The Legacy of Malthus: The Social Costs of the New Scientific Racism* (New York: Alfred A. Knopf, 1977).

Degler, Carl N. *In Search of Human Nature: The Decline and Revival of Darwinism in American Social Thought* (New York: Oxford University Press, 1991).

Eysenck, Hans J. *Race, Intelligence and Education* (London: Temple Smith, 1971).

———, *The Inequality of Man* (London: Temple Smith, 1973).

———, *Die Ungleichheit der Menschen* (Munich: Goldmann, 1978).

Fong, Melanie, and Larry O. Johnson. "The Eugenics Movement: Some Insight into the Institutionalization of Racism," *Issues in Criminology*, 9 (1974): 89–115.

Freeden, Michael. "Eugenics and Progressive Thought: a Study in Ideological Affinity," *Historical Journal*, 22 (1979): 645–71.

Gardener, James M. "Contribution of the German Cinema to the Nazi Euthanasia Program," *Mental Retardation*, 20 (1982): 174–75.

References

Gasman, Daniel. *The Scientific Origins of National Socialism: Social Darwinism in Ernst Haeckel and the German Monist League* (New York: American Elsevier, 1971).

Gibson, H. B. *Hans Eysenck: The Man and His Work* (London: Peter Owen, 1981).

Gottfredson, Linda S., and Jan H. Blits. "Employment Testing and Job Performance," *Public Interest*, 98, Winter (1990): 19–25.

Gordon, Robert A. "An Explicit Estimation of the Prevalence of Commitment to a Training School, to Age 18, by Race and by Sex," *Journal of the American Statistical Association*, 68 (1973): 547–53.

———, "Crime and Cognition: An Evolutionary Perspective." *Proceedings of the II International Symposium on Criminology*, vol. 4 (Sao Paulo: International Center for Biological and Medico-Forensic Criminology, 1975): 7–55.

———, "Examining Labeling Theory: The Case of Mental Retardation," *In the Labeling of Deviance: Evaluating a Perspective*, ed. Walter R. Gove (Beverly Hills: Sage, 1975): 83–146.

———, "Comment on 'Delinquency, Sex and Family Variables' by Andrew," *Social Biology*, 24 (1977): 337.

———, "Research on IQ, Race, and Delinquency: Taboo or Not Taboo?" *Taboos in Criminology*, ed. Edward Sagarin (Beverly Hills: Sage, 1980): 37–66.

———, "The Black–White Factor is g," *Behavioral and Brain Sciences*, 8 (1985): 229–31.

———, "Jensen's Contributions Concerning Test Bias: A Contextual View," *Arthur Jensen: Consensus and Controversy*, eds. Sohan Modgil and Celia Modgil (New York: Falmer Press, 1987): 77–154.

———, "SES versus IQ in the Race–IQ–Delinquency Model," *International Journal of Sociology and Social Policy*, 7.3 (1987): 30–96.

Graham, Loren R. "Science and Values: The Eugenics Movement in Germany and Russia in the 1920s," *American Historical Review*, 82 (1977): 1135–64.

Hahn Rafter, Nicole. *White Trash: The Eugenic Family Studies, 1877–1919* (Boston: Northeastern, 1988).

Haller, Mark H. *Eugenics: Hereditarian Attitudes in American Thought* (New Brunswick: Rutgers University Press, 1963).

Harwood, Jonathan. "The Reception of Morgan's Chromosome Theory in Germany: Inter-war Debate over Cytoplasmic Inheritance," *Medizinhistorisches Journal*, 19 (1984): 3–32.

———, "National Styles in Science: Genetics in Germany and the United States between the Wars," *ISIS*, 78 (1987): 390–414.

Hassencahl, Frances J. "Harry H. Laughlin: 'Expert Eugenics Agent' for the House Committee on Immigration and Naturalization, 1921 to 1931," diss., Case Western Reserve University, 1970.

Hellmann, Geoffrey. *Bankers, Bones & Beetles: The First Century of the American Museum of Natural History* (Garden City: The Natural History Press, 1969).

Higham, John. *Strangers in the Land: Patterns of American Nativism 1860–1924,* (New Brunswick: Rutgers University Press, 1963).

Hirsch, Jerry. "To 'Unfrock the Charlatans,'" *SAGE Race Relations Abstracts*, 6.2 (1981): 1–65.

———, and Barry Mehler. "Eugenics Has a Long Racist History," *Contemporary Psychology*, 31 (1986): 633.

Jensen, Arthur. "How Much Can We Boost I.Q. and Scholastic Achievement?," *Harvard Educational Review*, 39 (1969): 1–123.

———, *Genetics and Education* (New York: Harper & Row, 1972)

———, *Educability and Group Differences* (New York: Harper & Row, 1973).

———, "Rasse und Begabung," *Nation Europa*, 25.9 (1975): 19–28.

Kaiser, Jochen-Christoph. "Innere Mission und Rassenhygiene: Zur Diskussion im Centralausschuss für Innere Mission 1930–1938," *Lippische Mitteilungen*, 55 (1986): 197–217.

———, *Sozialer Protestantismus im 20. Jahrhundert: Beiträge zur Geschichte der Inneren Mission 1918–1934* (Munich: Oldenbourg, 1989).

Kelly, Alfred. *The Descent of Darwin: The Popularization of Darwinism in Germany, 1860–1914* (Chapel Hill: The University of North Carolina Press, 1981).

Kevles, Daniel J. *In the Name of Eugenics: Genetics and the Uses of Human Heredity* (Berkeley and Los Angeles: University of California Press, 1986).

Klee, Ernst. *"Euthanasie" im NS-Staat: Die "Vernichtung lebensunwerten Lebens"* (Frankfurt a.M.: Fischer, 1983).

Knorr-Cetina, Karin. *The Manufacture of Knowledge: An Essay on the Constructivist and Contextual Nature of Science* (New York: Pergamon Press, 1981).

Kranz, Harald. "Rassenhygiene/Eugenik in Deutschland: Institutionalisierung und Politisierung einer Wissenschaft (1927–1945)," thesis, University of Bielefeld, 1984.

Kroll, Jürgen. "Zur Entstehung einer naturwissenschaftlichen und sozialpolitischen Bewegung: Die Entwicklung der Eugenik/Rassenhygiene bis zum Jahre 1933," diss., University of Tübingen, 1983.

Kröner, Hans-Peter. "Die Eugenik in Deutschland von 1891–1934," diss., University of Münster, 1980.

Kühl, Stefan. *Bethel zwischen Anpassung und Widerstand: Die Auseinandersetzung der von Bodelschwinghschen Anstalten mit der Zwangssterilisation und den Kranken- und Behindertenmorden im Nationalsozialismus* (Bielefeld: AStA der Universität Bielefeld, 1990).

Lapon, Lanny. *Mass Murder in White Coats: Psychiatric Genocide in Nazi Germany and the United States* (Springfield: Psychiatric Genocide Research Institute, 1986).

Labisch, Alfons, and Florian Tennstedt. *Der Weg Zum "Gesetz über die Vereinheitlichung des Gesundheitswesens" vom 3. Juli 1934* (Düsseldorf: Akademie für öffentliches Gesundheitswesen, 1985).

Lagnado, Lucette Matalon, and Sheila Cohn Dekel, *Children of the Flames: Dr. Josef Mengele and the Untold Story of the Twins of Auschwitz* (London: Sidgwick & Jackson, 1991).

Lenk, Kurt. *"Volk und Staat": Strukturwandel politischer Ideologien im 19. und 20. Jahrhundert* (Stuttgart: Kohlhammer, 1971).

Lifton, Robert. *The Nazi Doctors: Medical Killing and the Psychology of Genocide* (New York: Basic Books, 1986).

Ludmerer, Kenneth M. "American Geneticists and the Eugenics Movement 1905–1935," *Journal of the History of Biology*, 2 (1969): 337–362.

———, *Genetics and American Society* (Baltimore and London: Johns Hopkins University Press, 1972).

References

Lutzhöft, Hans Jürgen. *Der Nordische Gedanke in Deutschland 1920 bis 1940* (Stuttgart: Klett-Cotta, 1971).

MacDowell, E. Charleton. "Charles Benedict Davenport, 1866–1944," *Bios*, 17 (1946): 3–50.

MacKenzie, Donald A. *Statistics in Britain 1865–1930: The Social Construction of Scientific Knowledge* (Edinburgh: Edinburgh University Press, 1981).

Mann, Gunter. *Biologismus im 19. Jahrhundert* (Stuttgart: Ferdinand Enke, 1973).

Marten, Heinz-Georg. *Sozialbiologismus: Biologische Grundpositionen der politischen Ideengeschichte* (Frankfurt and New York: Campus, 1983).

McLaren, Angus. *Our Own Master Race: Eugenics in Canada, 1885–1945* (Toronto, Ontario: McClelland & Steward, 1990).

Mehler, Barry. "John R. Commons," M.A. thesis, College of the City University of New York, 1972.

———, "The New Eugenics: Academic Racism in the U.S. Today," *Science for the People*, 15.3 (1983): 18–23.

———, "Eliminating the Inferior," *Science for the People*, 21.6 (1987): 14–18.

———, "A History of the American Eugenics Society, 1921–1940," diss., University of Illinois, 1988.

———, "Foundation for Fascism: The New Eugenics Movement in the United States," *Patterns of Prejudice*, 23.4 (1989): 17–25.

Miller, Marvin D. *Wunderlich's Salute: The Interrelationship of the German-American Bund, Camp Siegfried, Yaphank, Long Island, and the Young Siegfrieds and Their Relationship with American and Nazi Institutions* (New York: Malamud-Rose, 1983).

Mitscherlich, Alexander, and Fritz Mielke, *The Death Doctors* (London: Elek, 1962).

Mosse, George L. *The Crisis of German Ideology: Intellectual Origins of the Third Reich* (New York: Grosset and Dunlap, 1964).

———, *Toward the Final Solution: A History of European Racism* (London: J. M. Dent, 1978).

Mühlen, Patrick von. *Rassenideologien: Geschichte und Hintergründe* (Berlin and Bad Godesberg: J.H.W. Dietz, 1977).

Müller, Joachim. *Sterilisation und Gesetzgebung bis 1933* (Husum: Abhandlungen zur Geschichte der Medizin und der Naturwissenschaft, 1985).

Müller-Hill, Benno. *Die Tödliche Wissenschaft: Die Aussonderung von Juden, Zigeunern und Geisteskranken 1933–1945* (Reinbek bei Hamburg: Rowohlt, 1984).

Nachtsheim, Hans. "Die Frage der Sterilisation vom Standpunkt der Erbbiologen," *Berliner Gesndheitsblatt*, 1 (1950): 603–4.

———, *Für und Wider die Sterilisierung aus eugenischer Indikation* (Stuttgart: George Thieme, 1952).

———, "Das Gesetz zur Verhütung erbkranken Nachwuchses aus dem Jahre 1933 in heutiger Sicht," *Ärztliche Mitteilungen*, 33 (1962): 1640–44.

Neumärker, Klaus-Jürgen. *Karl Bonhoeffer: Leben und Werk eines deutschen Psychiaters und Neurologen in seiner Zeit* (Leipzig: Teubner, 1990).

Nowak, Kurt. *"Euthanasie" und Sterilisierung im "Dritten Reich": Die Konfrontation der evangelischen und katholischen Kirche mit dem "Gesetz zur Verhütung erbkranken Nachwuchses" und der "Euthanasie"-Aktion* (Halle: Niemeyer, 1977).

Nyiszli, Miklos. *Auschwitz: A Doctor's Eyewitness Account* (New York: Frederick Fell, 1960)

Paul, Diane. "Eugenics and the Left," *Journal of the History of Ideas,* 45 (1984): 567–90.

———, " 'Our Load of Mutations' Revisited," *Journal of the History of Biology,* 20 (1987): 321–35.

Pearson, Roger. *Introduction to Anthropology* (New York and London: Holt, Rinehart and Winston, 1974).

———, *Anthropological Glossary* (Malabar: Krieger Publishing, 1985).

———, *Race, Intelligence and Bias in Academe,* intro. Hans J. Eysenck (Washington: Scott-Townsend, 1991).

Pernick, Martin S. *The Black Stork: Eugenics and the Death of "Defective" Babies in American Medicine and Motion Pictures since 1915* (New York: Oxford University Press, 1992).

Pickens, Donald K. *Eugenics and the Progressives* (Nashville: Vanderbilt University Press, 1968).

Pogliano, Claudio. "Scienza e stirpe: Eugenica in Italia (1912–1939)," *Passato e Presente,* 5 (1984): 61–79.

Pois, Robert A. *National Socialism and the Religion of Nature* (London and Sydney: Croom Helm, 1986).

Poliakov, Léon, and Josef Wulf. *Das Dritte Reich und seine Denker* (Berlin: Arani, 1959).

Pollack, Michael. *Rassenwahn und Wissenschaft: Anthropologie, Biologie, Justiz und die nationalsozialistische Bevölkerungspolitik* (Frankfurt a.M.: Hain, 1990).

Proctor, Robert. *Racial Hygiene: Medicine under the Nazis* (Cambridge, London: Harvard University Press, 1988).

———, "Nazi Biomedical Technologies," *Lifeworld and Technology,* eds. Timothy Casey and Lester Embree (Washington: The Center for Advanced Research in Phenomenology and University Press of America, 1989): 17–48.

———, "Eugenics among Social Sciences: Hereditarian Thought in Germany and the United States," *The Estate of Social Knowledge,* eds. JoAnne Brown and David K. van Keuren (Baltimore: Johns Hopkins University Press, 1991): 175–208.

Provine, William B. "Geneticists and the Biology of Race Crossing," *Science,* 182 (1973): 790–96.

Race, Science and Society, ed. Leo Kuper (Paris: The UNESCO Press, 1975).

Reilly, Phillip R. "Involuntary Sterilization in the United States: A Surgical Solution," *The Quarterly Review of Biology,* 62 (1987): 153–70.

———, *The Surgical Solution: A History of Involuntary Sterilization in the United States* (Baltimore: Johns Hopkins University Press, 1991).

Roll-Hansen, Nils. "The Progress of Eugenics: Growth of Knowledge and Change in Ideology," *History of Science,* 26 (1988): 293–331.

———, "Geneticists and the Eugenics Movement in Scandinavia," *The British Journal for the History of Science,* 22 (1989): 335–46.

Rosenberg, Charles E. "Charles Benedict Davenport and the Beginning of Human Genetics," *Bulletin of History of Medicine,* 35 (1961): 266–76.

———, *No Other Gods: On Science and American Social Thought* (Baltimore: Johns Hopkins University Press, 1976).

References

Rost, Karl Ludwig. *Sterilisation und Euthanasie im Film des "Dritten Reiches"* (Husum: Matthiesen, 1987).

Rushton, J. Philippe. "Evolution, Altruism and Genetic Similarity Theory," *Mankind Quarterly*, 27 (1987): 379–95.

———, "Race Differences in Behavior: A Review and Evolutionary Analysis," *Personality and Individual Differences*, 9 (1988): 1009–24.

———, "Race Differences in Sexuality and Their Correlates: Another Look and Physiological Models," *Journal of Research in Personality*, 23 (1989): 35–54.

———, "The Reality of Racial Differences: A Rejoinder with New Evidence," *Personality of Individual Differences*, 9 (1989): 1035–40.

———, "Sir Francis Galton, Epigenetic Rules, Genetic Similarity Theory, and Human Life-History Analysis," *Journal of Personality*, 58 (1990): 117–40.

———, and Anthony F. Bogaert, "Population Differences in Susceptibility to AIDS: An Evolutionary Analysis," *Social Science and Medicine*, 28 (1989): 1211–20.

———, Charles Leslie, C. Owen Lovejoy, Glenn D. Wilson, and Peter J. M. McEwan, "Scientific Racism: Reflections on Peer Review, Science and Ideology," *Social Science and Medicine*, 31 (1990): 891–905.

Schmuhl, Hans Walter. *Rassenhygiene, Nationalsozialismus, Euthanasie: Von der Verhütung zur Vernichtung "lebensunwerten Lebens" 1890–1945* (Göttingen: Vandenhoeck & Ruprecht, 1987).

Schneider, William H. *Quality and Quantity: The Quest for Biological Regeneration in Twentieth-Century France* (Cambridge: Cambridge University Press, 1990).

Schreiber, Bernhard. *Die Männer hinter Hitler: Eine deutsche Warnung an die Welt* (Stuttgart: n.p., 1972).

Schwartz, Michael. "Sozialismus und Eugenik: Zur fälligen Revision eines Geschichtsbildes," *Internationale wissenschaftliche Korrespondenz zur Geschichte der deutschen Arbeiterbewegung*, 4 (1989): 465–89.

———, "Sozialistische Eugenik: Eugenische Sozialtechnologien in Diskurs und Politik der deutschen Sozialdemokratie, 1890–1933," diss., University of Münster, 1992.

Searle, Geoffrey R. *Eugenics and Politics in Britain, 1909–1914* (Leyden: Nordhoff International, 1976).

Segal, Lilli. *Die Hohenpriester der Vernichtung: Anthropologen, Mediziner und Psychiater als Wegbereiter von Selektion und Mord im Dritten Reich* (Berlin: Dietz, 1991).

Seidler, Horst. "Anthropologen im Widerstand?" *Der Widerstand gegen den Nationalsozialismus: Eine interdisziplinäre didaktische Konzeption zu seiner Erschliessung*, ed. Maria Zenner (Bochum: Brockmeyer, 1989): 67–122.

Shapiro, Thomas M. *Population Control Politics: Women, Sterilization and Reproductive Choice* (Philadelphia: Temple University Press, 1985).

Smith, John David. *Minds Made Feeble: The Myth and Legacy of the Kallikaks* (Rockeville: Aspen Systems Corporation, 1985).

Stepan, Nancy. *The Idea of Race in Science: Great Britain 1800–1960* (London: Macmillian, 1982)

———, *The Hour of Eugenics: Race, Gender and Nation in Latin America* (Ithaca: Cornell University Press, 1991).

Thomann, Klaus-Dieter. "Otmar Freiherr von Verschuer-ein Hauptvertreter der faschistischen Rassenhygiene," *Medizin im Faschismus*, eds. Achim Thom and Horst Spaar (Berlin: VEB Verlag Volk und Gesundheit, 1985): 36–52.

Trombley, Stephen. *The Right to Reproduce: A History of Coercive Sterilization* (London: Weidenfeld & Nicolson, 1988).

Turner, Henry A. *Die Grossunternehmer und der Aufstieg Hitlers* (Berlin: Siedler, 1985).

Wagener, Otto. *Hitler aus nächster Nähe: Aufzeichnungen eines Vertrauten 1929–1932*, ed. Henry A. Turner (Frankfurt a.M.: Ullstein, 1978).

Weber, Matthias M. "Die Entwicklung der Deutschen Forschungsanstalt für Psychiatrie in München zwischen 1917 und 1945," *Sudhoffs Archiv*, 75 (1991): 74–89.

Weindling, Paul. "Soziale Hygiene: Eugenik und medizinische Praxis—Der Fall Alfred Grotjahn," *Das Argument: Jahrbuch für kritische Medizin* (1984): 6–20.

———, "Weimar Eugenics: The Kaiser Wilhelm Institute for Anthropology, Human Heredity, and Eugenics in Social Context," *Annals of Science*, 42 (1985): 303–18.

———, "Fascism and Population Policies in Comparative European Perspective," *Population, Resources and the Environment: The Interplay of Science, Ideology and Intellectual Traditions*, eds. M. Teitelbaum and J. Winter (Cambridge: Cambridge University Press, 1988): 102–20.

———, "From Philanthropy to International Science Policy: The Rockefeller Funding of Biomedical Sciences in Germany, 1920–1940," *Science, Politics and the Public Good: Essays in Honor of Margret Gowing*, ed. Nicolaas Rupke (Basingstoke: Macmillan, 1988): 119–40.

———, *Health, Race and German Politics between National Unification and Nazism, 1870–1945* (Cambridge: Cambridge University Press, 1989).

———, "The 'Sonderweg' of German Eugenics: Nationalism and Scientific Internationalism," *The British Journal of the History of Science*, 22 (1989), 321–33.

Weingart, Peter. "German Eugenics between Science and Politics," *Osiris*, n.s., 5 (1989): 260–82.

———, Jürgen Kroll, and Kurt Bayertz, *Rasse, Blut und Gene: Geschichte der Eugenik und Rassenhygiene in Deutschland* (Frankfurt a.M.: Suhrkamp, 1988).

Weinreich, Max. *Hitler's Professors: The Part of Scholarship in Germany's Crimes against the Jewish People* (New York: Yiddish Scientific Institute, 1946).

Weiss, Sheila F. *Race Hygiene and National Efficiency: The Eugenics of Wilhelm Schallmayer* (Berkeley, Los Angles, London: University of California Press, 1987).

———, "The Race Hygiene Movement in Germany, 1904–1945," *The Wellborn Science: Eugenics in Germany, France, Brazil and Russia*, ed. Mark B. Adams (New York and Oxford: Oxford University Press, 1990): 8–68.

Index

Abel, Karl, 59
abortions, 30
Ackermann, Heinrich. *See* Günther, Hans F. K.
Acquired Immune Deficiency Syndrome (AIDS), 7–8
Allen, Garland, xiv
American Association for the Advancement of Science (AAAS), 7, 66
American Committee on Maternal Health, 42
American Eugenics Society: conferences of, 73, 76; during and after World War II, 100, 106; founder of, 86; membership of, 66, 74; Northern California Branch, 57; relation to Nazi Germany, xiv–xv, 45–46, 73, 76, 95; Southern California Branch, 45; transformation of, 79–84
American Genetic Association, 18, 57, 66
American Immigration Control Federation, 9–10
American Journal of Public Health and The Nation's Health, 54
American Museum of Natural History, 67, 86
American Psychological Association, 111 n.14
American Public Health Association, 45, 54
American Social Hygiene Association, 66
American Society for Human Genetics, 103
American Sociological Association, 66

Angriff, Der, 35
Annals of Eugenics, 105
Annals of Human Genetics, 105
anti-Semitism, 66, 73, 75–76, 78, 97–99. *See also* Jews
Archiv für Rassen- und Gesellschafts-biologie (ARGB), 18, 24, 38
Asians, presumed differences from whites and blacks, 3–4, 7–8, 106
Association for Voluntary Sterilization, 101
Astel, Karl, 28, 30, 33
asylums, xiii, 40, 48–49, 56, 101, 107 n.1
Auschwitz, 60, 102–3
Austrian League for Regeneration and Heredity, 25

Babbot, Frank, 66, 128 n.5
Bajema, Carl, xiv
Baker, Lewellys F., 67
Barkan, Elazar, 80
Bauer, Julius, 69
Baur, Erwin, 19, 50, 71
Bayertz, Kurt, xvi
Bell, Alexander Graham, 14–15
Berliner Börsenzeitung, 34, 99
Besselmann, Paul Heinz, 38
Bethel Institution, xiii, 107 n.1
Bigelow, Maurice A., 100
Binet, Alfred, 40
birth control, 78
Birthright, Inc., 105–6
Blackmar, Frank W., 40
blacks: and the American eugenics movement, 75; presumed differences

Printed in the USA
CPSIA information can be obtained
at www.ICGtesting.com
CBHW071005310724
12461CB00002B/4